Principles of Wildlife Conservation

Principles of Wildlife Conservation

Dr. Ankur Awadhiya, IFS

CRC Press
Taylor & Francis Group
Boca Raton London New York

CRC Press is an imprint of the
Taylor & Francis Group, an **informa** business

First edition published 2022
by CRC Press
6000 Broken Sound Parkway NW, Suite 300, Boca Raton, FL 33487-2742

and by CRC Press
2 Park Square, Milton Park, Abingdon, Oxon, OX14 4RN

Library of Congress Cataloging-in-Publication Data

Names: Awadhiya, Ankur, 1987- author.
Title: Principles of wildlife conservation / Dr. Ankur Awadhiya, IFS.
Description: Boca Raton : CRC Press, 2021. | Includes bibliographical
  references and index.
Identifiers: LCCN 2020058429 | ISBN 9781032023328 (hardback) | ISBN
  9780367479930 (paperback) | ISBN 9781003037545 (ebook)
Subjects: LCSH: Wildlife conservation.
Classification: LCC QL82 .A93 2021 | DDC 333.95/416--dc23
LC record available at https://lccn.loc.gov/2020058429

ISBN: 978-0-367-47993-0 (pbk)
ISBN: 978-1-032-02332-8 (hbk)
ISBN: 978-1-003-03754-5 (ebk)

DOI: 10.1201/9781003037545

Typeset in Nimbus font
by KnowledgeWorks Global Ltd.

To Ushi

Contents

Preface

The 1960s are considered a watershed period in our thinking about the environment. For time immemorial, the focus of humankind had been to conquer nature in the name of 'progress' — converting forests into settlements, prairies into farmlands and ranches, rivers into dammed sources of electricity and so on. In this process, we strived to vanquish natural processes, quell, quash, overpower and overcome Mother Nature — and bring her to knees. And we did all this while preparing for our own destruction — through atomic weapons, wars and rampant abuse of our own surroundings. But this was to change soon after — particularly in the '60s.

In 1956, we discovered the Minamata disease — a neurological condition caused by mercury poisoning — that was rapidly affecting people in Japan — with extremely tragic human costs. When scientists dwelled on the genesis of the disease, they found that a chemical corporation was dumping untreated industrial waste — used catalysts — into the sea. And these wastes contained mercury. The company could have easily stopped the dumping — or at least treated the wastes before dumping — but that entailed costs, and so was not done. Perhaps we had become such egoists through generations of abusing nature that we just did not care! The bottom line was the top priority — and the life of sea-dwelling plants and animals was immaterial. But what goes around, comes around. The sea organisms exposed to the mercury compounds took up those mercury compounds into their bodies, and bio-magnified the concentrations up the food chain — with the result that local fishes came to have some of the highest concentrations of mercury ever found in an organism. Fishermen caught these sea fishes, sold them in the market and all those who ate tainted fish got exposed to mercury poisoning! Humans realised that trashing the environment is not without consequences. Karma *is* a bitch.

In the year 1962, Rachel Louise Carson, an American marine biologist, published a book — *Silent Spring* — in which she detailed the impacts of overuse of pesticides in our farmlands. Pesticides were one of the foundations of the Green revolution, and the book presented a direct — dark and ghastly — link between the use of DDT and the extermination of bird populations. The name of the book is derived from the fact that springs were getting silenced — so few birds remained that their chirps, chitters, tweets, trills and songs, especially during the spring season, became conspicuously absent. Before this book, Green revolution was only an emancipator — a saviour to the hungry millions. But now we needed to contemplate the costs at which this food security was being pulled off. The Aral Sea, once the fourth largest lake in the world, had also started to dry since the incoming waters were being diverted to cotton fields.

Around this time, trepidations about the future of our planet, and the destiny of humankind in particular, began to be felt. In 1968, Dr. Paul Ehrlich, in his book — *The Population Bomb* — predicted that rising populations, if left uncontrolled, would thrust humanity to its annihilation. We needed to change, if we wanted to survive. We began to use computers to model our destiny. The World3 computer simulation developed at MIT, and the 1972 publication of the Meadows book — *The Limits to Growth* — showed that if we do not mend our ways — quickly and decisively — our future is going to be immensely grim. In many of the simulations, the world order collapsed due to pollution or depletion of resources. Photographs showing the Earth from space — particularly 'Earthrise' by the Apollo 8 crew in 1968, and 'The Blue Marble' by the Apollo 17 crew in 1972, demonstrated that our planet is fragile — and that nature is not something to be subverted. We began pushing for reforms. The Earth Days were first celebrated in 1970, and Greenpeace was founded in 1971. The United Nations Conference on the Human Environment — held at Stockholm in 1972 — resulted in a declaration on 26 principles concerning the environment and development, an action plan and a resolution. International and national laws were passed to protect the environment, our planet and the species that live on our planet. We became 'concerned citizens' of this planet.

So what has changed in the last 50 years — since the 1970s? Well, our population doubled — from around 4 billion in 1970 to nearly 8 billion in 2020. We became more 'advanced' — our consumption of resources, particularly electricity and fossil fuels, escalated like never before. The ensuing global warming and climate change meant that these are no longer topics of mere academic interest — we now hear them even in presidential elections. Habitats shrunk — to make way for agriculture, houses and other infrastructures. We used groundwater to near exhaustion. Habitats got fragmented when 'developmental projects' criss-crossed all lands. Extinctions rocketed, and we wiped out wildlife populations by as much as 60% — from their 1970s levels!

All this is not without consequences. Each decade is warmer than the previous one — and more and more people are dying of heat strokes than ever before. Areas such as Australia are witnessing long spells of droughts — killing ranches and farmlands. The city of Cape Town almost completely ran out of water. Diseases are wiping out olive trees in Europe, increased temperatures are eliminating coffee plantations in Africa and South America. California wildfires are turning more and more violent and destructive — impacting numerous communities, both near and far. The pollutants reaching our food and waters have ensured that more and more of us — even young children — fall prey to diseases, particularly metabolic and hormonal diseases. Encroachment into wildlife habitats has guaranteed easy access to zoonotic diseases — and our marvellous modes of transport have greatly empowered diseases like COVID-19 to turn them into pandemics. Our coasts are under a constant threat of submergence; climatic extremes are leading to more intense cyclones and storms; our agriculture is failing, owing to salinisation and water scarcity; our dams are failing, due to sedimentation; and even after so much 'progress' and 'development,' today we are less happy, less contented and more diseased — than we were in the 1970s.

Why is this so? Do we have a dearth of people who care for the environment? I don't think so. Data suggests that the number of caring individuals is actually on the rise. After all, we did not have movements like the school strike for climate in the 1970s. Our presidents and vice presidents did not talk much about climate and the environment in the 1970s.

In that case, do we have a dearth of scientific knowledge about what needs to change? Well, scientific knowledge has all but increased — year-after-year. Today we have a much better understanding of the science of destruction, greatly superior mathematical models and extremely powerful computers to simulate them. We have profoundly refined our predictive capabilities through advancements in science and technology.

But we do lack a commitment to the cause — perhaps we ourselves are not confident enough. We lack fortitude — to say no to decisions that harm the environment. Perhaps we lack even the obligatory eloquence needed to put forth our views and concerns to the policy and decision makers.

A lot of this has to do with our public lack of understanding. We love wildlife, we love the environment, but we often do not know how to conserve them. Such topics are seldom, if ever, taught in schools. Often the solutions are simple, but they're simply inaccessible to the right audience. We can save habitats from fragmentation by roads by constructing underpasses — but the Civil Engineering syllabus of most universities does not incorporate and discuss underpasses at all. We can solve the problems of floods and droughts by preserving wetlands — but most policy makers are only taught about the need of laying pipelines for water, not how to bring water to those pipelines.

Wildlife reserves can bring employment to locals and raise the GDP sustainably, but the politicians are only presented with the options of developing farming or industry to benefit communities — never the option to create and expand wildlife reserves. This is true even when the land is barren enough to be useless for agriculture and so far off that it is uneconomical for any industry!

This needs to change. And the present book is an attempt in this direction. I've tried to explain the concepts and principles of wildlife conservation, as I've understood through my academic studies, observations during visitations and learnings from practical experiences in the field. This book lays bare the workings of the ecosystem, the threats to wildlife and the what's and how's of saving them. The book is heavy on pictures, to enable the reader to see as I've seen, and feel as I've felt. I am sure they'll be found useful.

I wish to thank the faculty and friends at IIT Kanpur, the Indira Gandhi National Forest Academy and the Wildlife Institute of India, together with the officers in my department who were kind enough to share their insights and experiences about Biology and Conservation. My mother, despite her illness and old age, meticulously scrutinised the manuscript, and provided several constructive suggestions. Words are insufficient to express my profound sense of gratitude to her and to my family. Thank you all.

Dr. Ankur Awadhiya, IFS

Contributors

Dr. Ankur Awadhiya (b. 1987) is an IFS officer of 2014 batch borne on the Madhya Pradesh cadre. Trained as an engineer, he earned his B. Tech in Biological Sciences and Bioengineering in 2009 from IIT Kanpur, followed by a Ph.D from IIT Kanpur in 2015, AIGNFA (Honours Diploma) from IGNFA Dehradun in 2016 and Post Graduate Diploma in Advanced Wildlife Management (Honours Diploma) from WII Dehradun in 2018. He maintains a keen interest in academics and research and has been a recipient of several honours including the NTSE scholarship, KVPY fellowship, Shri P. Srinivas Memorial prize, K. P. Sagriya Shreshta Vaniki puraskar and the S. K. Seth Prize. Besides academics, he maintains a passion towards photography, painting, movie making and creative literary pursuits.

CHAPTER 1

Introduction

1.1 WHAT IS WILDLIFE?

We all have an instinctive understanding of wildlife. Wildlife is often perceived in terms of the reference examples that come to our mind. Often wildlife is something like a tiger, a ferocious, dangerous animal. Or something like a panda, a cuddly creature. If we peruse a dictionary, we would find that the word 'wild' is defined as something that is 'not domesticated or cultivated.' In other words, something found in, or living in the natural environment, typically forests. And the word 'life' denotes a capability of demonstrating aspects like growth, reproduction and activity that differentiate the living from the dead.

Thus 'wildlife' refers to those living things (plants and animals) that are undomesticated, or are living in, the natural environment [Fig. 1.1 a–c]. But this definition may lead us to certain queer directions. This is because often some wild animals are domesticated and put to human use. A good example is the elephant that has been used for ages for military conquests, and also as a working beast to carry large logs in the forests. What happens when an elephant is domesticated? Does it cease to be wildlife? What about when an elephant calf is in the *process* of being domesticated [Fig. 1.1d], but has not yet become domesticated? Can we draw a line for the 'level' of domestication that would make an animal cease to be a wildlife? What about a bear that was used in a circus, but now has been rescued, and is being re-wilded [Fig. 1.1e]. When does it become 'wild' again? Where do we draw a line? If a tiger strays into an agricultural field (an artificial environment), will it cease to be wildlife? What about a tiger in a zoo [Fig. 1.1f]? These questions make us think whether it is actually possible to draw a line between the *natural environment* and the *artificial environment* at all. In certain cases when a forest is being encroached upon and getting converted into a farmland, or a farmland that is being laid fallow and trees are coming up on it, there may be numerous small pockets where forest and farmland meet. In these circumstances, there may not even be a *line* to distinguish the *natural environment* from the *artificial environment*.

When we consider the management of wildlife, we need to understand wildlife, and define wildlife in concrete terms. Definitions are important because they help us understand the nuances of what we are attempting to do. Of course, we cannot manage wildlife without knowing what we are trying to manage! The Oxford Dictionary of English [Stevenson, 2010] defines wildlife as:

"wild animals collectively; the native fauna (and sometimes flora) of a region."

But what is 'native'? How long does one species have to remain in a place to qualify as a native species? *Lantana camara* is a species that was introduced to India over a century ago as an ornamental plant to be used in gardens. Over time, it has spread outside the gardens; it is an invasive species. Today large pockets of forests have dense growth of this plant. It not only provides shelter, but also food in the form of flowers and fruits to several species of animals and birds [Fig. 1.2]. Many species are now dependent on lantana, and if we try to exterminate this species, there will be large ecological impacts. So now, should we count *Lantana camara* as a 'native' species, and call

(a) Sal trees in a forest.

(b) Chital in Ranthambhore Tiger Reserve.

(c) Parakeet in Sariska Tiger Reserve.

(d) Elephant calf being trained in Mudumalai Tiger Reserve.

(e) A bear in a bear rescue facility, in the process of being re-wilded.

(f) A tiger in Chhatbir Zoo.

Figure 1.1: Some examples of wildlife. Undomesticated animals living in natural environments (a–c) are easily recognised as wildlife. However, it is difficult to draw a line in case of domesticated animals, or animals living in man-made environments (d–f).

it wildlife? Or look at its history and still call it is a non-native species? And if we choose to call it non-native, will we still call it a non-native species after a century? What about after a millennium? After all, most species at any place have arrived at some time in the past. Where do we draw a line between 'native' and 'non-native'? And even more importantly, do we *need* to draw a line? Can't we just say that the because the species that are present today in the wild conditions play a role in the ecosystem, they are wildlife and need to be protected and managed?

These questions can be debated upon *ad infinitum* for academic purposes. But wildlife management cannot wait that long. So for the purposes of this book, we consider a broad definition of wildlife, one that permits us to include large number of species. It is the legal definition of wildlife. The Wildlife (Protection) Act 1972 defines wildlife as:

"wild life includes any animal, aquatic or land vegetation which forms part of any habitat."

Note the word *includes*. This means that "any animal, aquatic or land vegetation which forms part of any habitat" is wildlife, but wildlife may also comprise somethings that are not "any animal, aquatic or land vegetation which forms part of any habitat." In other words, the definition leaves scope for incorporating other forms of 'life,' say microorganisms, into the scope of 'wildlife.' This is important because our understanding of wildlife and the relationships between different forms of life, their importance and the threats being faced by them is constantly evolving. As we imbibe greater details of the Biology and Ecology, we can incorporate more organisms into our list of 'wildlife'. As per this definition, wildlife includes not only animals, but also plants. And these animals and plants may form part of *any habitat*, even *artificial habitat* such as farmlands or cities. These animals and plants may be 'wild' in nature (such as a tiger in the forest), 'domesticated' in nature (such as a trained elephant being used in forestry operations), or anywhere in between.

1.2   WHAT IS CONSERVATION?

The word 'conservation' is derived from the Latin prefix *con* (meaning *together*) attached to the Latin word *servare* (meaning *keep*). Thus,

Conservation ■ Keeping together.

When we conserve something, we keep it together, protect it, and do not let it break down. Conservation may be defined as the "advocacy or practice of the sensible and careful use of natural resources." It includes things such as sustainable harvesting (preventing overuse of resources and removing only that much as will permit the resource to continue, or be *kept together* for a very long period of time), wise use of soil (avoiding soil erosion, maintenance of fertility, etc.) and water (avoiding overuse, pollution, and so on).

Wildlife conservation emphasises advocacy and practice of sensible, careful and wise use of wildlife resources in a manner that they remain *kept together* for a very long time. It allows non-consumptive utilisation of wildlife resources through sustainable tourism, and may even permit sustainable harvesting of wildlife resources through controlled poaching, but in a manner that does not harm the resource *in toto*.

Wildlife conservation is different from 'wildlife preservation' which emphasises "allowing some places and some creatures to exist without significant human interference." Thus, wildlife preservation emphasises that certain wildlife (and areas bearing them) must be left to themselves and not be used at all, whereas wildlife conservation permits sustainable usage of wildlife and the areas bearing them. However, both are also related since excessive usage of wildlife resources may warrant that we leave certain patches for the system to recover, much like leaving a farmland fallow to regain fertility. Thus, wildlife preservation may be considered a component of wildlife conservation, but wildlife conservation has a much broader scope than wildlife preservation.

(a) Dense growth of lantana in a forest.

(b) Lantana supports several insects such as butterflies.

(c) A parakeet feeding on lantana fruits and seeds in Sariska Tiger Reserve.

Figure 1.2: What is native may not be clearly distinguishable.

We also differentiate conservation from 'environmentalism' which is "concern about the impact of people on environmental quality." In environmentalism, the emphasis is on environmental quality, not on wildlife resources. However, here again, both are related since much of the impetus to conserve wildlife has come from the understanding that maintenance of wildlife habitats is crucial for maintenance of environmental quality. This is because of the 'ecosystem services' provided by wildlife habitats in the form of air and water purification, prevention of soil erosion, etc. Thus while environmentalism provides a motive for wildlife conservation and wildlife conservation helps improving the environmental quality, the scope and focus of wildlife conservation is very different from environmentalism.

Wildlife lives in a wild environment, its habitat. It is dependent on its habitat for its requirements such as food, water and shelter. Thus when we aim to manage wildlife, we need to have a good understanding of its habitat and the management of the habitat as well. The field of wildlife conservation is largely based upon this understanding of the relationship between organisms and their environment, something that is studied in the science of Ecology. We may even state that conservation of wildlife is in some sense "Applied Ecology," or the application of the knowledge gained from Ecology for the conservation of wildlife. An understanding of the environment in which wildlife lives and prospers is thus essential to conserving wildlife; and in this book, we shall explore several aspects of Ecology as a foundation for Wildlife Conservation.

1.3 THE NEED FOR CONSERVING WILDLIFE

Why should we care about conserving wildlife? Isn't it more prudent to just care about ourselves? Is something wrong in thinking, "every animal to itself?"

There are several ways of thinking about this issue. One approach considers that all the animals and plants that exist today were made by God, the Almighty. And if He ordained that these organisms should exist, then they should. We must not go against His desires. And given that over the ages we humans have unleashed massive destructions to numerous organisms, now is the time to 'atone the sins' — reduce the sins through conserving what is left. However, one difficulty with this line of thinking is that we do not know if He wanted us to destroy the organisms in the first place! Because after all, many species have gone extinct even without our interference.

Another approach considers that we humans are just one of the several species on this planet. There is nothing special about us humans. And all the species have equal 'right to life.' And thus, we need to conserve wildlife because we have, through several foolish actions, brought innumerable sufferings and destruction to several species. However, one difficulty with this line of thinking is that one may even argue that there is no such 'right to life' for all the species. After all, rights are fought for and had, not presented on a platter. And also that we humans are special for we are the 'masters of the earth and all that lives in it.' Thus, we are within our rights to use each and every organism as we deem fit. There is nothing wrong with destroying organisms.

One way out of this dilemma is to consider our best interests, not those of the wildlife. Our thinkers and philosophers have often pondered over near versus future benefits. One example comes from the fable of the goose that laid golden eggs. The farmer who owned the goose got greedy, and in an attempt to have all the gold at once, killed the bird. However, opening the bird, he found that there was no gold inside. Thus while on the one hand, the farmer could have had golden eggs once every few days for many years to come, on the other hand, he lost out on all the gold due to his greed. This story is pertinent because wildlife also provide us with certain benefits, time and again. If we conserve them, we shall have those benefits for several years — nay, several generations. But if we try to have all the benefits at once, we will have no benefits. This line of thinking states that we must conserve wildlife to have the benefits that the wildlife offer to us.

These benefits include revenue and income through tourism, wildlife products such as honey and timber, and so on. Thus, from a purely profit point of view, if we wish to maximise the benefits, we will have to use the resources in a managed manner. If we wish to have the benefits for a very long time, the maintenance of adequate wildlife stocks is essential, which is another way of saying that wildlife will need to be *kept together*, or conserved. Hence we need to understand that *wildlife is a natural resource*, where resource is defined as "a stock or supply of economic and productive factors that can be drawn on to accomplish an activity." Wildlife is a natural resource since it exists without actions of humans.

Classification of natural resources

Natural resources can be classified as follows:

1. On the basis of origin

    (a) biotic resources: coming from living matter e.g. timber

    (b) abiotic resources: coming from non-living matter e.g. iron ore

2. On the basis of stage of development

    (a) potential resources: those resources that may be used in the future e.g. oil that has not been drilled

    (b) actual resources: those resources that are currently being used after surveying, quantification and qualification e.g. timber from forest

    (c) reserve resources: the part of actual resources that can be developed profitably in the future e.g. low concentration ores

    (d) stock resources: those resources that have been surveyed but we lack the technology to use them e.g. hydrogen for nuclear fusion

3. On the basis of renewability

    (a) renewable resources: those resources that can be replenished naturally e.g. sunlight

    (b) non-renewable resources: those resources that either form slowly or do not naturally form at all in the environment e.g. coal

---

Wildlife resources are biotic, renewable, actual (sometimes potential) resources. Thinking in this manner, we can easily calculate the costs and benefits of doing any activity that concerns wildlife. Suppose a project to set up mining in a wildlife habitat costs USD 5 million, provides a benefit of USD 10 million and harms the wildlife benefits worth USD 8 million, it is easy to see that the costs of the activity (USD 5 million to set up the mine + USD 8 million in the form of benefits foregone = USD 13 million total) are greater than the benefits that the activity will provide (USD 10 million). And so no rational thinker will agree to the project; only a simpleton may agree to go ahead with this activity.

Note how this line of thinking underlines the importance of knowing the *value* of wildlife. If we did not (or could not) compute and consider the value of the benefits foregone (USD 8 million), we would have actually gone ahead with the project, since the apparent benefits (USD 10 million) are greater than the apparent costs (USD 5 million). It is of utmost importance to understand the economic valuation of wildlife resources by valuing the various services being offered by them [Fig 1.3]. These comprise of:

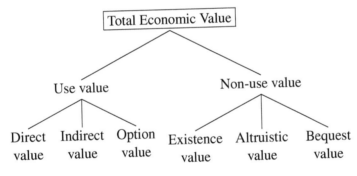

Figure 1.3: Components of total economic value.

1. Use value: Value arising out of use of the resource. This includes:

    (a) Direct value: Direct value comprises of:

        i. consumptive and productive values such as

            A. timber

            B. firewood

            C. medicines

            D. grazing

            E. other non-timber forest produce

            F. water, etc.

        ii. non-consumptive values such as

            A. recreation — ecotourism

            B. education (for students of Ecology, Taxonomy, Systematics, etc.) and research (as natural experimental areas for Ecology, store-house of medicinal plants, etc.)

         C. places of human (e.g. several tribal and forest-dependent communities) and wildlife habitat, etc.

   (b) Indirect value: These include

     i. watershed benefits, including

       A. agricultural productivity

       B. soil conservation

       C. ground water recharge

       D. regulation of stream flows

     ii. ecosystem services, such as

       A. nitrogen fixation

       B. waste assimilation

       C. air and water purification

       D. carbon sequestration and storage

       E. microclimatic functions and microclimate regulation

     iii. evolutionary processes, including

       A. global life support

       B. biodiversity

   (c) Option value: It is an option for the future direct and indirect use of the forest. That is to say, it values the future use of forests by keeping the present forests intact and sustainably utilised.

2. Non-use value: Value arising even though the resource is not being used. This includes:

   (a) Existence value: This is the value deriving from the knowledge that the resources continue to exist. For instance, we Indians derive pleasure value from knowing that polar bears exist, even though we may never directly use them.

   (b) Altruistic value: This is the value derived from the knowledge of use of resources by others in the current generation. For instance, we in Madhya Pradesh derive value from knowing that the forests of Assam are of use of the people of Assam, our compatriots, even when those forests are not directly being used by us in Madhya Pradesh.

   (c) Bequest value: This is the value of leaving use and non-use values for offsprings or future generations. For instance, we value conserving tigers for our children and grandchildren.

We'll explore actual valuation of wildlife resources in detail in Chapter 11. For now it is evident that if we wish to have these wildlife benefits, we'll need to conserve wildlife.

However, the current business as usual (BAU) scenario poses several threats to wildlife resources, including poaching and over-exploitation of wildlife (think killing of the goose that laid golden eggs), changes in climate due to global warming, and destruction and loss of wildlife habitats due to deforestation, mining, agricultural expansion, and other reasons that we explore in Chapter 4. Conservation of wildlife requires dedicated efforts to minimise these threats and to reverse the damages already done.

1.4 THE ORGANISATION OF THIS BOOK

This book will cover both fundamentals and applied aspects of conserving wildlife. In Chapter 2, we'll look at organisation of the living world, biodiversity, energetics and interactions between different organisms. An understanding of these forms the foundation of wildlife management, and helps guide the management of habitats. Chapter 3 will cover Population Ecology, Community Ecology and Animal Behaviour. These aspects aid in taking control of low populations of species of interest. Threats to wildlife resources are dealt in detail in Chapter 4, and monitoring of wildlife numbers is discussed in Chapter 5. Monitoring is a crucial facet of wildlife management, since it is extremely difficult to manage something if we don't have a quantitative way of knowing if our management interventions are having any desired impacts. Without a quantitative valuation, we are merely groping in the dark. We'll examine habitats and their management in Chapter 9. Since habitats are places where organisms live, a better habitat often results in higher population densities of target species. Chapter 8 concerns wildlife genetics, which is especially crucial when we have small populations to manage. In the absence of scientific management, breeding between close relatives may result in a genetic collapse of the species — even local extinctions — and that needs to be prevented at all costs. Two macro modes of wildlife conservation, the *in-situ* and the *ex-situ* management are detailed in Chapter 10. Chapter 6 examines the principles of managing wildlife diseases, with case studies on managing specific wildlife diseases. This is followed by animal restraint and immobilisation in Chapter 7. We shall review the emerging aspects of wildlife management in Chapter 11, including the use of new technologies of remote sensing, geographic information systems, drones and topics in conservation economics.

CHAPTER 2

Organisation of life

2.1 WHAT IS ORGANISATION?

Organisation is the act of arranging (something) systematically to efficiently coordinate (its) activities. We see organisation all around us, whether it is the organisation of clothes in different drawers, or arrangement of books in the shelves of a library, or the division of an office into different sections. We organise because it is efficient. It takes less time to navigate the closet to look for a favourite pair of socks, we can get to the desired book quickly and it is easy to find and reach the concerned officer for a given issue. In a very similar manner, wildlife too is organised — lions have prides, bisons live in herds, and birds flock together. This arrangement helps them organise their activities efficiently.

We can explore organisation by understanding two underlying principles of organisation:

1. The Hierarchical Principle : "Hierarchy emerges inevitably because hierarchical structures are more stable."

2. The Emergent Principle: "The whole has properties that its parts do not have."

2.1.1 The hierarchical principle

In his classic paper [Herbert et al., 1962], Herbert Simon narrates a story of two watchmakers, Hora and Tempus. Both make very fine watches, and are constantly in demand, with new customers frequently calling them. However, with time, Hora's prosperity increases whereas Tempus becomes poorer and poorer, and ultimately has to close his shop.

The reason? Watches are complex devices with around 1,000 parts. The watches Tempus made were so designed that a partly assembled watch was unstable, and fell to its constituent pieces whenever he had to answer a call. Then the watch had to be reassembled from scratch. The more customers called Tempus, the less uninterrupted time he had to finish a single watch.

On the other hand, Hora's watches were designed in a hierarchical fashion. There were subassemblies of around 10 elements each that were stable and did not fall down to constituent pieces when left on their own. These subassemblies could be put together into larger subassemblies, and so on till the final watch got made. Thus, any interruption would only result in the loss of a small part of Hora's work, comprising roughly 10 elements. This resulted in Hora maintaining a much greater efficiency than Tempus.

This story describes how hierarchy leads to efficiency. We can observe applications of similar hierarchical organisations in most complex systems. For example, an army is organised as follows:

$$Squad \rightarrow Section \rightarrow Platoon \rightarrow Company \rightarrow Battalion$$

$$\rightarrow Regiment \rightarrow Brigade \rightarrow Legion \rightarrow Corps \rightarrow Army$$

Each of these subassemblies is a stable unit: it has rules and procedures, and knows how to perform in a given circumstance. Each subassembly is in turn comprised of several smaller sub-assemblies. Organisation such as this is what makes modern army such a formidable force. It permits quick and efficient transfer of information so that the desired results are obtained without many errors.

Similarly in nature as well, hierarchy "emerges almost inevitably through a wide variety of evolutionary processes, for the simple reason that hierarchical structures are stable [Chen and Chie, 2007]." Many sub-cellular organelles come together to form a cell. Many cells form a tissue. Many tissues form an organ. The list goes on [Table 2.1]:

Sub-cellular organelle → Cell → Tissue → Organ → Organ system → Organism → Population → Community → Ecosystem → Biome → Biosphere

Table 2.1: Levels of organisation in the biological world.

Level	Description
Sub-cellular organelle [Fig. 2.1a]	"A specialised subunit within a cell that has a specific function" e.g. mitochondrion, chloroplast, nucleus, vacuole
Cell [Fig. 2.1a]	"The basic structural, functional, and biological unit of all known living organisms, the smallest unit of life"
Tissue [Fig. 2.1a]	"An ensemble of similar cells and their extracellular matrix from the same origin that together carry out a specific function"
Organ [Fig. 2.1b]	"Collections of tissues with similar functions"
Organ system [Fig. 2.1b]	"A group of organs that work together to perform one or more functions"
Organism [Fig. 2.1c]	"An individual entity that exhibits the properties of life"
Population [Fig. 2.1d]	"All the organisms of the same group or species, which live in a particular geographical area, and have the capability of interbreeding"
Community [Fig. 2.1e]	"A group or association of populations of two or more different species occupying the same geographical area and in a particular time"
Ecosystem [Fig. 2.1f]	"A community made up of living organisms and nonliving components such as air, water, and mineral soil"
Biome	"A community of plants and animals that have common characteristics for the environment they exist in"
Biosphere	"The worldwide sum of all ecosystems"

2.1.2 The emergent principle

As we move up the hierarchy, several new properties emerge that were not present in the lower rungs of the hierarchy. An examples is the conversion of individual wolves into a pack. A wolf can run after a prey till it gets exhausted. Thus, there is a good chance that a prey escapes the wolf. However organisation of wolves into a pack brings several emergent properties. In a pack, the wolves not only run after their prey, they also communicate amongst themselves about how best to catch the prey. Through communication (an emergent property since it requires two or more wolves) the pack can work as a relay. Through communication, the wolves arrange themselves in

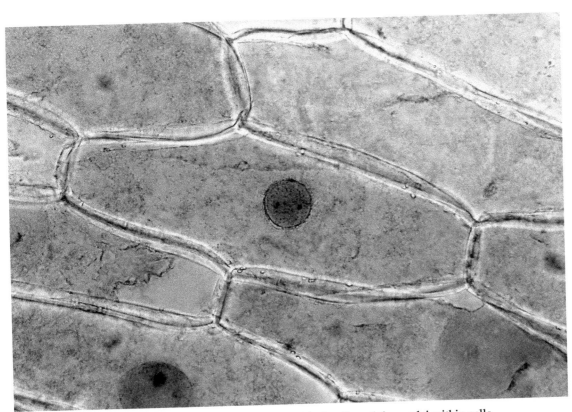

(a) Onion epidermis tissue showing individual cells and the nuclei within cells.

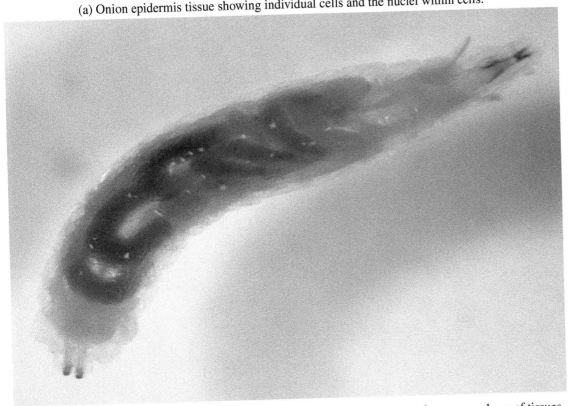

(b) The digestive system of *Drosophila melanogaster* larva comprises of several organs made up of tissues.

(c) An organism such as this leopard in Gir National Park has several organ systems.

(d) This flamingo population in Johannesburg comprises several individual organisms.

(e) The community has several populations. In Manas Tiger Reserve, we can observe rhinoceros population interacting with hog deer population.

(f) The ecosystem in Manas Tiger Reserve has living as well as non-living constituents.

Figure 2.1: Examples of organisation in the biological world.

such a manner that the exhausted wolf is replaced by another — and another — and another — till the prey gets exhausted and gets killed by the pack. This property of communication was absent at the level of an individual wolf — there isn't much need for an organism to communicate with itself. However as we move up the hierarchy, communication becomes important and acts as a force multiplier. Similarly the property of cooperation emerges when we move from the level of individual to the level of the pack. Group dynamics emerge, as does competition.

We can observe new properties emerging at all levels of organisation [Table 2.1]. When cells are organised into tissue, they form bonds with each other. Communication happens through chemical and electrical signals. And phenomena such as apoptosis (programmed cell death) emerge that were absent at the level of individual cells. Since these emergent properties increase efficiency, they get selected through the process of evolution.

We can visualise these emergent properties in several ways. When migrating birds form flocks, they move in formations. The flight formations of birds often appear as V-shaped or U-shaped [Fig. 2.2a] to take advantage of a phenomenon known as *upwash*. The upward moving air — formed by wingtip vortices due to wing movement of another bird — is captured and assists each bird — except the leader at the tip of the 'V' — to support its own weight in flight. It is as if air moving upward is pushing the bird upwards, countering gravity to an extent. This is similar to the action of other upward-moving air currents such as convection currents, which permit birds to glide for long durations without flapping their wings. This reduces fatigue and conserves energy. The 'leader' at the tip of the 'V' does not get any advantage of upwashes, since there is no bird ahead of it. So there is a chance that it gets tired. The birds have a solution — the leader gets rotated in a timely fashion, and as the leader begins to feel fatigued, another bird comes to the front to take its place, and the previous leader moves back in the 'V'. This spreads the flight fatigue equally among all members of the flock. This property of using upwash — which is also used by fighter jets flying in formation — has emerged naturally in flocks of birds and is an emergent property.

Another emergent property is herd behaviour [Fig. 2.2b]. When animals in herds are attacked by predators, they move in a fashion that confuses the predator. This gives all members of the herd a survival advantage. This collective movement is an emergent property. We also observe this with schools of fishes.

We humans also make use of emergent properties. When houses are organised and arranged to form a city, it gives rise to several emergent properties [Fig. 2.2c]. In every city, roads are laid out in a definite order. Houses are arranged in a certain fashion. This is done to maximise the efficiency of reaching any particular place in minimum time and cost. If the roads are at 90°, it also helps in collecting garbage which is moved by winds to specific spots. This is one reason why even during the time of the Harappan civilisation, the cities had roads cutting at right angles. A very similar phenomenon is also observed in ant hills and termite mounds [Fig. 2.2d]. The organisation of a termite mound is such that air currents are taken advantage of. They cool and dry the mound, keeping the interiors in amicable conditions. This is an emergent property that arises due to organisation.

The following two statements describe the Emergent Principle:

1. "The whole is greater than the sum of the parts."

2. "The whole has properties its parts do not have."

2.2 BIODIVERSITY

Organisation and hierarchy have permitted nature to evolve several forms. Since small changes do not collapse the complete system, they are tolerated. This permits variations in populations. Thus in a population, there will be individuals with slightly different characteristics. Some individuals

(a) Birds flying in V-shaped formation.

(b) A herd of impalas in Kruger National Park is organised to protect itself from predators. The predators (cheetahs) are between the impalas and the vehicle in the background.

(c) The city of Jodhpur has properties that individual houses don't have.

(d) Similarly, a termite mound shows several emergent properties.

Figure 2.2: Examples of the Emergent Principle.

will have better than average eyesight, some will have an extremely good sense of smell, some will be able to run faster than others, and so on. And depending on the situation at hand, one or more of these characteristics will provide survival advantages to the individuals that have them. Say in a season of drought, the food available can be best discovered through smell. In that case, the individuals with a better-than-average sense of smell will be able to get food easily, and will survive and reproduce, such that the next generation will have more individuals with a better sense of smell. On the other hand, if the food could be discovered through colour and not smell, the individuals with a better-than-average sense of sight will be able to get food easily, and will survive and reproduce, such that the next generation will have more individuals with a better sense of sight. Thus the variety of forms plays a big role in ensuring the survival of the population. Even in desperate times, there will be some individuals that will survive, and the species will continue.

Consider, on the other hand, a population where every individual is the same — a clone of each other. In that population, if one individual gets sick, there is a good chance that the disease will equally be able to infect all others, and may even wipe off the population. In a sense, variations are a key to survival. This is why we find in nature a variety of life in all forms — whether bacteria, fungi, plants, animals — and at all levels of organisation — from genetic to ecosystem-level. This variety of life is known as biodiversity. And this biodiversity provides resilience in the face of changes. Even at an ecosystem level, if one prey is wiped off, the availability of other preys helps the predator to survive. If we had only one prey and one predator, the loss of the prey would guarantee the extinction of the predator as well.

2.2.1 Biodiversity at genetic level

We take some of our traits from our father, and some others from our mother. Genetics is the science of how these traits are passed through heredity. The unit of heredity that gets transmitted from parents to offsprings is known as 'gene.' Thus we have genes for eye colour, hair colour, hair shape, height, diseases like diabetes, and so on.

Genetic biodiversity is the diversity of this genetic information. This diversity of genes can be measured at the levels of phyla, families, species, populations and individuals.

Measures of genetic biodiversity include

1. Polymorphism (P), which is the proportion or percentage of genes that are polymorphic (occurring in several different forms such that no allele has a frequency of >95%). In essence, it is like asking how many traits have variations. Do we have variations only in the hair colour in a population, or do we also see variations in the texture, size, density, etc. The more the fraction of genes that are polymorphic, the more genetically biodiverse the population is.

2. Heterozygosity (H), which is the proportion or percentage of genes for which the average individual is heterozygous. In other words, the proportion of genes for which the two copies (one received from each parent) in the average individual are different. More heterozygosity is an indicator of free mating amongst the population, whereas less heterozygosity indicates that mating is constrained by distance, relationship or other factors. More heterozygosity often corresponds to greater genetic biodiversity and resilience in the population.

2.2.2 Biodiversity at species level

Species are groups of actually or potentially interbreeding natural populations, which are repro-ductively isolated from other such groups [Mayr, 1942]. Thus, even though a tiger from India will probably never mate with a tiger from Siberia, they comprise one species (*Panthera tigris*) since they are not reproductively isolated. This means that they can *potentially* interbreed, even if they are actually not interbreeding. On the other hand, tigers and elephants living close together are reproductively isolated. So they neither actually interbreed amongst each other, nor can *potentially* interbreed. Thus they form different species (*Panthera tigris* vs. *Elephas maximus*).

Any area can have fewer or more number of species. Species biodiversity asks the question: How many species are there, and how are they distributed? The answer indicates the levels of

1. *Species richness*, or the number of species that are present. More the number of species, more is the species biodiversity. The number of species present can be computed using a species accumulation curve [Fig. 2.3], which plots the number of species discovered against the effort spent in discovering those species. To understand, we may consider a person visiting a new area in the search of species. Every day the person spends 8 hours to search for species. On day 1, she finds 40 different species. On day 2, she notes 50 species. But of these, 35 were also seen in the previous day. Thus the number of new species discovered on day 2 is 50 − 35 = 15. And the total number of species found till day 2 is 40 (new species on day 1) + 15 (new species on day 2) = 55. On day 3, she finds 45 species, of which 7 were also seen on day 1 (and not on day 2), 20 were seen on day 2 (and not on day 1), and 8 were seen on both day 1 and 2. Thus out of the 45 species seen on day 3, a total of 35 species are those that were seen before. And only 10 new species were found on day 3. Thus, the total number of species discovered till day 3 = 55 + 10 = 65.

We can observe that while the number of species discovered goes on increasing with the effort exerted, the rate of increase (or the marginal increase) goes on decreasing (40 → 15 → 10…). Ultimately an asymptote is reached where newer species are no longer found even with extended effort. This asymptote gives the number of species in the area under study.

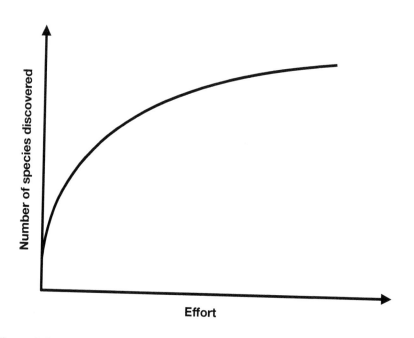

Figure 2.3: Species accumulation curve as a measure of species richness.

2. *Species evenness*, or the distribution of individuals of different species. More the evenness, more is the species biodiversity. To understand, consider two communities with following distribution of individuals:

Number of individuals of different species in community 1 — A: 90, B: 2, C: 3, D: 5

Number of individuals of different species in community 2 — A: 30, B: 20, C: 35, D: 15

Even though both the communities have four species (A, B, C and D), community 2 has more biodiversity because in any portion of the community, the probability of finding all four species is much greater than in community 1, which is dominated by a single species (A). This will also reflect in the greater resilience that community 2 will have over community 1 in the face of changes.

Two indices that combine the information about species riches and species evenness are the Simpson's index and the Shannon's index.

2.2.2.1 Simpson's diversity index

$$D = \frac{1}{\Sigma_{i=1}^{S} P_i^2}$$

$$E = \frac{D}{D_{max}}$$

where
D is the Simpson's diversity index
S is the number of species in the area
P_i is the proportion of the ith species
E is the index of equitability or evenness
D_{max} is the maximum possible value of Simpson's diversity index, given by S (D is maximum when each species is represented by 1 and only 1 individual)

2.2.2.2 Shannon's diversity index

$$H = -\Sigma_{i=1}^{S} P_i ln P_i$$

$$J = \frac{H}{H_{max}}$$

where
H is the Shannon's diversity index
S is the number of species in the area
P_i is the proportion of the ith species
J is the index of equitability or evenness
H_{max} is the maximum possible value of Shannon's diversity index, given by lnS

2.2.3 Biodiversity at ecosystem level

Ecosystem biodiversity asks the question: How many ecosystems are there, and how are they distributed? Thus a forest with woodland, grassland, lakes, rivers and cliffs will have a greater biodiversity than another forest with only woodland. Since different ecosystems and ecotones support different species, a large ecosystem-level biodiversity often translates to a large species-level biodiversity as well.

2.2.4 Spatial component of biodiversity

The spatial component of biodiversity asks the question: How is the biodiversity distributed in an area? It can be described as:

1. α biodiversity: the diversity that exists within an ecosystem

2. β biodiversity: the diversity that exists among different ecosystems

3. γ biodiversity: the diversity that exists among different geographies

This is important since measures proposed to increase biodiversity at one scale may reduce biodiversity at other scales. To understand, consider the example shown in figure 2.4. The first panel represents an island with two different ecosystems: a woodland and a swamp. The woodland has trees in which two species of lizards thrive. The swamp has grasses and supports one species of lizards.

Suppose a proposal suggests that the swamp should be drained and replaced with a woodland, so that two species of lizards can be supported in place of one, in the area occupied by the swamp. This proposed situation is represented in the second panel. We can observe that on the patch that originally had the swamp, there are now two species of lizards in place of one. So the *biodiversity has increased on the local scale*.

But we also observe that the total number of species on the island now is two in place of three. Thus, the *biodiversity has decreased on the larger scale*.

In other words, while the α biodiversity has increased from one to two, the β biodiversity has reduced from three to two.

This has important ramifications for conservation. Often we wish to have large biodiversity, since it provides a degree of resilience to changes. This is especially important in the current era of global warming and climate changes. Nowadays a large amount of habitat manipulation (often referred to as *habitat enhancement*) is being proposed. An example is the conversion of grasslands and scrub forests into woodlots, often using monoculture of hardy species such as teak (*Tectona grandis*). This is very often highly counter-productive, since it not only reduces biodiversity on larger scales (Consider what would happen if all the patches of grasslands — often called *natural blanks* — were to be planted with trees. Where would the species living on these *natural blanks* go? And if they become extinct, will it not reduce biodiversity and resilience of the ecosystem?), but often the interventions fail miserably (referred to as *foresters' folly*, since the planted species often do not survive in the *natural blanks*). And hence any proposal purporting to *increase biodiversity* through interventions must be examined with a lump of salt. An analysis would reveal that often nature knows best in such matters, and no intervention is the best intervention.

2.2.5 Why do some areas have more biodiversity, and some have less?

Not all places have the same amounts of biodiversity. Some, such as equatorial rainforests and coral reefs have very high biodiversity, while some others such as deserts and permafrosts have less biodiversity. Yet others such as agricultural fields with monoculture (only one crop being cultivated) and using chemicals such as fertilisers, insecticides, pesticides and weedicides have extremely less biodiversity. Those areas with large biodiversity that needs conservation are known as 'biodiversity hotspots,' defined as areas with

1. a high species richness having a large number of species per unit area such that any loss of this area will impact several species

2. a high degree of endemism: having a number of species that are only found in these areas and nowhere else such that a loss of this area will translate into the extinction of these species

3. a high degree of threat that, if unchecked, will lead to large losses of biodiversity very soon: implying an urgency of protection that is needed for this area.

A map of the biodiversity hotspots of the world is depicted in figure 2.5.

But why are only these areas the biodiversity hotspots? Why don't all areas have high biodiversity? Why do we have some areas with large biodiversity, and some others with smaller biodiversity? There are several factors at play. In particular, we have five main hypotheses that are used to explain the different levels of biodiversity in different areas:

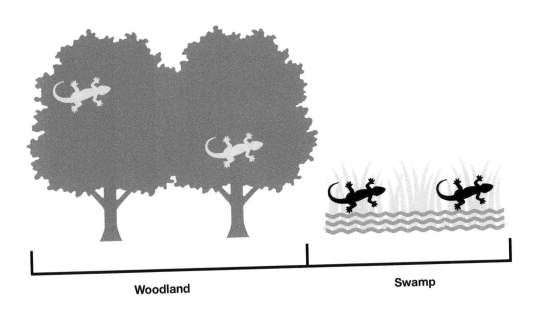

(a) A piece of woodland with two species of lizards, and a swamp with one species.

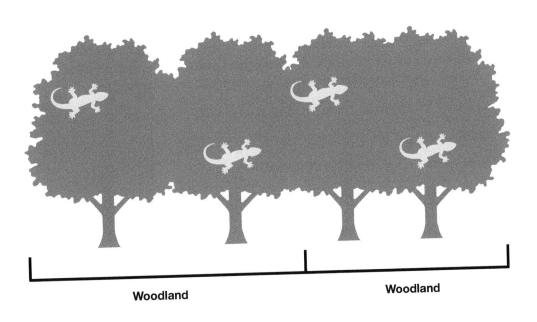

(b) When the swamp is drained and converted into woodland, the α biodiversity increases, but the β biodiversity decreases.

Figure 2.4: An example to depict the importance of measuring biodiversity.

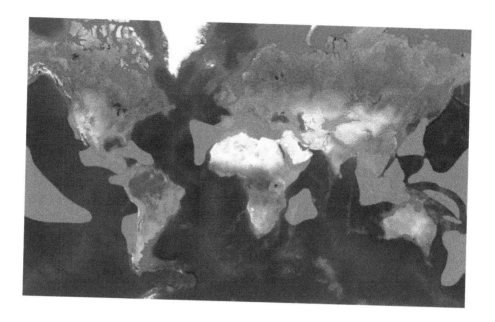

Figure 2.5: A map depicting the biodiversity hotspots of the world.

1. Evolutionary speed hypothesis: It works on the premise that biodiversity requires time to build up. Thus, there is more biodiversity in areas that have had

   (a) more time to evolve — such that older areas such as old forests will have more biodiversity than newer areas such as recently formed islands, and / or

   (b) more rapid evolution — such that if the species have shorter generation times and higher mutation rates and the area is such that natural selection is acting more quickly, say, due to high competition, then we will have more biodiversity in that area.

2. Geographical area hypothesis: It states that there is more biodiversity in areas with

   (a) larger size, and / or

   (b) physically or biologically complex habitats,

   for these can support more number of organisms and more niches (or roles such as herbivory or scavenging) for organisms, which translates into more number of species. In other words, there is more area, and more kinds of situations to occupy by different species adapted to these situations, which will result in more species being present. At the same time, larger number of organisms in the larger area also acts as a buffer against extinction due to catastrophic events, meaning that the species, once formed, will continue to exist for long periods of time, increasing the biodiversity of the area.

3. Interspecific interactions hypothesis: It states that there is more biodiversity in areas with

   (a) more competition — which affects niche partitioning and differentiation, permitting development of more specialised species, and

   (b) more predation, which retards competitive exclusion and promotes faster action of natural selection, which translates into evolution of more number of species, faster.

4. Ambient energy hypothesis: It states that there is more biodiversity in areas with more ambient energy. This is because energy is required for the functioning of organisms. At the same time, the other extreme with very large ambient energy will again support a smaller number of species, since very few species will be able to tolerate the climatically unfavourable very hot conditions. This hypothesis explains why the polar ice caps — with very little ambient energy — have very little biodiversity in comparison to equatorial forests — which have larger ambient energy, and why deserts — with very hot climates — have little biodiversity in comparison to more equable climates.

5. Intermediate disturbance hypothesis: It states that there is less biodiversity in areas with

   (a) very high / frequent disturbances, which leads to extinction of species, and

   (b) very low / infrequent disturbances, which leads to competitive equilibrium and slower action of natural selection process, resulting in lesser number of new species evolving over time.

   Thus, this hypothesis posits that areas with an intermediate level of disturbance should have larger biodiversity than areas with very high or very low levels of disturbance. This hypothesis also explains why human-dominated areas such as cities and agricultural fields — with very high and frequent anthropogenic disturbances — have very less biodiversity.

2.3   WILDLIFE ENERGETICS

Energy is the ability to do work. All organisms require energy to carry on their life processes. The branch of science that studies energy is known as energetics. It deals with "the properties of energy and the way in which it is redistributed in physical, chemical, or biological processes." Ecological Energetics is the discipline that concerns itself with how energy (and matter) moves through different organisms in an organised manner, through food chains and webs.

Food chains [Fig. 2.6a] are defined as "the linear flow of energy and matter in an ecosystem." It is a representation of "who eats whom." When an organism (called 'predator') feeds on another organism (called 'prey'), the matter and energy in the body of the prey moves into the body of the predator. The food chain begins at a source of matter and energy, typically plants which make their own food through the process of photosynthesis. Later, food and energy get transferred from their source in plants, through herbivores and carnivores to higher-order carnivores, and finally to detritivores (waste or debris feeders) or decomposers that release the energy and matter back into the environment. When different food chains interact, they form a food web [Fig. 2.6b].

Depending on whether they can make their own food, or have to depend on others for food, organisms can be divided into two categories:

1. Autotrophs (*auto* = self, *troph* = nourishment): These are the organisms that make their own food using light or chemicals. They are responsible for primary production. Autotrophs are also known as primary producers, and include organisms such as trees, plants and algae. Autotrophs are further divided into:

   (a) Photoautotrophs (*Photo* = light, *auto* = self, *troph* = nourishment): Photoautotrophs are organisms that use light as a source of energy to manufacture organic molecules and food. e.g. most plants.

   (b) Chemoautotrophs (*Chemo* = chemical, *auto* = self, *troph* = nourishment): Chemoautotrophs are organisms that use chemical reactions as a source of energy to manufacture organic molecules and food. e.g. *Hydrogenovibrio crunogenus*, bacteria found in deep-sea hydrothermal vents.

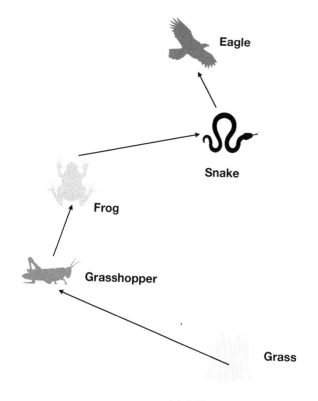

(a) A linear food chain.

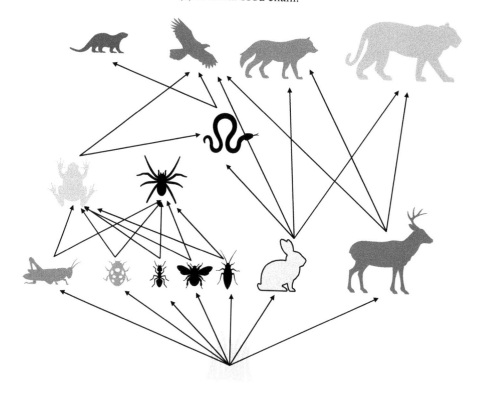

(b) A food web consisting of several interacting linear food chains.

Figure 2.6: Food chain and food web.

2. Heterotrophs (*hetero* = other, *troph* = nourishment): These are organisms that cannot produce their own food, and so must depend on other organisms for their food. Examples include most animals, including us humans.

In ecology, we differentiate organisms into producers and consumers:

1. Producer: An organism that makes its own food. e.g. autotrophs

2. Consumer: An organism that consumes some other organism for food. e.g. heterotrophs

Consumers are of several kinds, including

1. Primary consumer: An organism that only consumes the producers. e.g. grasshopper

2. Secondary consumer: An organism that consumes the primary consumer. e.g. frog

3. Tertiary consumer: An organism that consumes the secondary consumer. e.g. snake

4. Quaternary consumer: An organism that consumes the tertiary consumer. e.g. hawk

Depending on their source of food, consumers can also be classified into

1. Herbivores: Organisms that eat only plants. They are primary consumers. e.g. cow

2. Carnivores: Organisms that eat other animals. They are secondary, tertiary or quaternary consumers. e.g. tiger

3. Omnivores: Organisms that eat both plants and other animals. They are generally secondary or tertiary consumers. e.g. bear

4. Decomposers: Organisms that convert dead material into soil and recycle nutrients.
   Decomposers include detritivores and microorganisms. Detritivores consume detritus (decomposing plant and animal parts as well as faeces), and make it more exposed to the action of microbial decomposers such as bacteria and fungi that further break it down into simple compounds.

2.3.1   Different kinds of food chains

Depending on whether they start from a plant base, or from detritus, food chains are classified into

1. Grazing food chains: These start from a plant base and move through herbivores to carnivores. Grazing food chains can be:

    (a) predator food chains
        e.g. Grass → Chital → Tiger
        The size of organisms generally increases as we move up the chain.

    (b) parasite food chains
        e.g. Rat → Flea → Parasitic protozoa
        The size of organisms generally decreases as we move up the chain.

2. Detritus food chains: They start from detritus, and move through detritivores to carnivores.
    e.g. Fallen leaves of mangroves → Detritivores → Detritivore consumers e.g. small fish or insect larvae → Small fish → Large fish → Piscivorous birds

The differences between grazing and detritus food chains are listed in table 2.2.

Table 2.2: Differences between grazing and detritus food chains.

Characteristic	Grazing food chain	Detritus food chain
Primary source of energy	Sun	Detritus
First trophic level	Herbivores	Detritivores
Length	Generally long chains	Generally shorter chains

2.3.2 Food webs and ecological pyramids

A food web is a system of interlocking and interdependent food chains [Fig. 2.6b]. When these food chains interact, they create *trophic levels*, each of the several hierarchical levels in an ecosystem, that consist of organisms sharing the same function in the food chain (e.g. producers, consumers or decomposers) and the same nutritional relationship to the primary sources of energy. Thus in a forest food web, all species of plants, whether herbs, shrubs or trees form the trophic level of primary producers. All the primary consumers, whether they are insects, birds or animals that eat the plants form the trophic level of primary consumers, and so on. The integration of different organisms into trophic levels helps us to analyse the working of the ecosystem and the food webs, through ecological pyramids.

2.3.2.1 Ecological pyramids

Ecological pyramids are graphical representations designed to show the energy, biomass or number of organisms at each trophic level in a given ecosystem. They are also known as trophic pyramids, Eltonian pyramids, energy pyramids and food pyramids. They are depicted as a pyramidal shape with a large base that tapers towards a vertex.

At the base of an ecological pyramid is the trophic level of producers, on top of which is the trophic level of herbivores, followed by carnivores, and so on till the apex predator. Often a trophic level is represented as a rectangle, the width of which represents the energy, biomass, or number at the given trophic level.

In a majority of cases, the pyramids have a broad base and taper towards the top, since the energy, biomass, or number at a lower trophic level is greater than that at a higher trophic level. This is because as they move through a food web, there is a loss of matter and energy at each trophic level. For example, when a cow eats 10 kg of grass, its weight does not increase by 10 kg when the grass is digested and assimilated. A large chunk of the matter in the grass is egested out with faeces, and a majority of the energy in the grass is used for bodily processes in the cow, including temperature regulation, movement, respiration and flow of blood. Only a small portion of the matter and energy that were in the 10 kg of grass get stored in the cow's body. Similarly, there is a loss of matter and energy at each trophic level.

There are three kinds of ecological pyramids:

1. Pyramids of energy: These represent the energy contained in all the organisms at each trophic level. The base represents the amount of energy contained in all the organisms at the producer trophic level. The next rung represents the amount of energy contained in all the primary consumers (herbivores), and so on. Pyramid of energy is always upright [Fig. 2.7a] since there is a loss of energy from each trophic level to the next.

2. Pyramids of biomass: These represent the sum of biomass of all the organisms at each trophic level. The base represents the amount of biomass contained in all the organisms at the producer trophic level. The next rung represents the amount of biomass contained in all the primary consumers (herbivores), and so on.

Pyramid of biomass is generally upright [Fig. 2.7a]. But in certain cases it can be inverted [Fig. 2.7b], especially when the lower rungs of the pyramid are comprised of organisms that have high births and deaths. A good example is found in marine ecology. Planktons are small microscopic organisms found in water that are used as food by many species of fish. These microscopic organisms can multiply at a very fast rate, but they have a short life span. At any point of time, there will be a population of planktons, but with each organism being microscopic, the total biomass of the planktons will be less. However, with their fast rates of multiplication, they are almost always available as food for the fishes. The fishes that eat the planktons have a larger body mass. But they too are much smaller than their predators, and make up for it by having a fast rate of reproduction. The total biomass at their trophic level is smaller than at the trophic level of their predators, which are large-sized. Such a scenario generates an inverted pyramid of biomass.

3. Pyramids of numbers: These represent the numbers of organisms at each trophic level. The base represents the number of all the organisms at the producer trophic level. The next rung represents the number of all the primary consumers (herbivores), and so on.

Pyramid of numbers can be upright [Fig. 2.7a], as in a grassland ecosystem. The number of grass plants is much greater than the number of herbivores, which is much greater than the number of carnivores, and so on. But we also have inverted pyramid of numbers [Fig. 2.7c], such as in a tree ecosystem. A large tree such as a banyan tree can support hundreds of birds. Each bird can support hundreds of parasites such as ticks and fleas. And each parasite can support numerous hyperparasites such as bacteria and protozoa. This make the pyramid of numbers inverted.

We can also have spindle-shaped pyramids of numbers where the middle rungs are much larger than either the top or the bottom. A grove can support a much greater number of frugivorous (fruit-eating) birds than the number of fruit trees. But the large number of frugivorous birds will only support a small number of predators such as hawks. This makes for a spindle-shaped pyramid of numbers [Fig. 2.7d]. Another example is the marine ecosystem, where the number of zooplanktons can be much larger than the number of phytoplanktons that they eat, and the number of fishes that eat the zooplanktons [Fig. 2.7e].

We also have dumb-bell shaped pyramid of numbers where the middle rung is smaller than the rungs above and below. An example is a grassland where a large number of grass plants support a small number of rabbits. But each rabbit can support a large number of parasites such as fleas, making the pyramid of numbers dumb-bell shaped [Fig. 2.7f].

2.3.2.2 Efficiency of energy transfer

When an organism eats food, the matter and energy in the food get utilised in several ways. A portion of the food may be digested, and some portion may be egested with faeces as undigestible matter. Of the portion that gets digested and absorbed, a part will be used to meet the requirements of the organism. The organism needs energy to meet its requirements of locomotion, feeding, bodily processes, thermoregulation, respiration, etc. During respiration the food is burnt inside the cells in the presence of oxygen to generate carbon dioxide and water, and energy is released that can be utilised by the organism. In this process, matter in the form of carbon dioxide and water vapour is released into, and lost to, the environment. Only a small portion of the ingested matter and energy get incorporated into the body of the organism in the form of new cells, fat storage or glycogen storage. The efficiency of energy transfer investigates the efficiency of each of these processes.

We begin by defining the standing crop. A standing crop is the total dried biomass of the living

(a) Upright pyramid of energy, biomass, or numbers.

(b) An inverted pyramid of biomass.

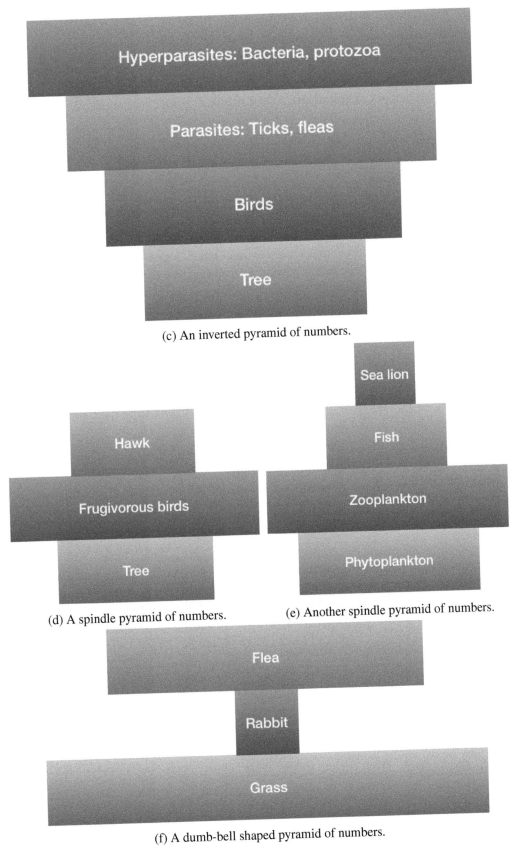

(c) An inverted pyramid of numbers.

(d) A spindle pyramid of numbers.   (e) Another spindle pyramid of numbers.

(f) A dumb-bell shaped pyramid of numbers.

Figure 2.7: Examples of ecological pyramids.

organisms present at a trophic level. This gives us an estimate of the matter and energy at each trophic level. The efficiency is defined and computed as follows:

1. Ecological efficiency: The efficiency with which energy is transferred from one trophic level to the next.

2. Exploitation efficiency: The amount of food ingested divided by the amount of prey production:

$$\text{Exploitation efficiency} = \frac{I_n}{P_{n-1}}$$

where n is the nth trophic level and $n-1$ is the $(n-1)^{th}$ trophic level, or the preceding trophic level. I denotes ingestion and P denotes production.

3. Assimilation efficiency: The amount of assimilation (integration and modification of food molecules after the processes of digestion and absorption) divided by the amount of food ingestion:

$$\text{Assimilation efficiency} = \frac{A_n}{I_n}$$

where A denotes assimilation.

4. Gross production efficiency: Consumer production divided by amount of ingestion:

$$\text{Gross production efficiency} = \frac{P_n}{I_n}$$

5. Net production efficiency: Consumer production divided by amount of assimilation:

$$\text{Net production efficiency} = \frac{P_n}{A_n}$$

6. Ecological efficiency: Ecological efficiency is the exploitation efficiency multiplied by the assimilation efficiency multiplied by the net production efficiency, which is equivalent to the amount of consumer production divided by the amount of prey production:

$$\text{Ecological efficiency, EE} = \frac{I_n}{P_{n-1}} \times \frac{A_n}{I_n} \times \frac{P_n}{A_n} = \frac{P_n}{P_{n-1}}$$

2.3.2.3 The 10 percent rule

When ecological efficiency was computed for several ecosystems, it was found that during the transfer of energy from one trophic level to the next, only about 10% of the energy gets stored as biomass, while the remaining is either

1. lost during transfer, say, when the prey runs away from the predator, both using their stored energies,

2. lost due to incomplete digestion, since not everything is digestible and is digested by the predator and a portion gets lost as faeces, or

3. used for bodily functions and reproduction of the predator.

At the same time, the efficiency of plants in capturing the Sun's energy through photosynthesis is only around 1%.

Thus, in the food chain:

Grass → Grasshopper → Frog → Snake → Hawk

if 100,000 Joules of energy from the Sun was intercepted by the grass, the amount of energy assimilated at each stage would be:

100,000 Joules from Sun → Grass (1,000 Joules, 1%) → Grasshopper (100 Joules, 10%) → Frog (10 Joules, 10%) → Snake (1 Joule, 10%) → Hawk (0.1 Joule, 10%)

Considering the plants as level 1, we have

$$\text{Energy at } n^{th} \text{ level} = \frac{\text{Energy intercepted from the Sun}}{10^{n+1}}$$

Since the amount of energy available at each successive trophic level decreases, not many organisms can be supported at higher trophic levels. This explains why we do not find very long food chains in nature.

2.3.3 Primary production

Primary production — or production done by producers — is the synthesis of organic compounds from atmospheric or aqueous carbon dioxide, through the processes of photosynthesis or chemosynthesis. Photosynthesis not only converts the ultimate source of energy (the Sun) to bioenergy, fuelling the complete ecosystem, but also releases oxygen as a by-product — to be used by several organisms during respiration. Since primary producers produce their own food and are not dependent upon other organisms for food, they are also known as autotrophs. Examples include trees, plants and algae.

Two processes happen in tandem in primary producers:

1. Photosynthesis / chemosynthesis, e.g.

$$6CO_2 + 6H_2O \xrightarrow[\text{Solar energy}]{\text{Chlorophyll, enzymes}} C_6H_{12}O_6 + 6O_2$$

2. Respiration, e.g.

$$C_6H_{12}O_6 + 6O_2 \xrightarrow{\text{Metabolic enzymes}} 6CO_2 + 6H_2O$$

During photosynthesis / chemosynthesis, food is produced using the energy from the Sun or from the environment. During respiration, food is consumed to provide energy for bodily processes. While photosynthesis releases oxygen and consumes carbon dioxide, respiration consumes oxygen and releases carbon dioxide. In plants, photosynthesis occurs during the day time when sufficient light is available. However, respiration continues day and night. Thus during the day, $Photosynthesis > Respiration$, and during the night, $Respiration > Photosynthesis$. There are times when $Photosynthesis = Respiration$, often during early mornings and late evenings. This equilibrium point for plants where photosynthesis equals respiration is known as the **Compensation point**.

2.3.3.1 Energetics of primary production: Terms

To analyse the energetics of primary production, we define the following terms:

1. Gross primary production Energy (or carbon) fixed via photosynthesis.

2. Net primary production: Gross primary production − Energy (or carbon) lost via respiration.

3. Productivity: Production per unit time:

$$Productivity = \frac{Production}{Time}$$

4. Gross primary productivity: Gross primary production per unit time.

5. Net primary productivity: Net primary production per unit time.

It can also be represented as:

$Net\ primary\ productivity = APAR \times LUE$

where

- APAR = Absorbed photosynthetically active radiation (MJ / m^2 / time)
- LUE = Light use efficiency (grams carbon per MJ energy)

6. Efficiency of gross primary production: Fraction of energy in incident sunlight that gets fixed by gross primary production:

$$\eta_{gross} = \frac{Energy\ fixed\ by\ gross\ primary\ production}{Energy\ in\ incident\ sunlight}$$

7. Efficiency of net primary production: Fraction of energy in incident sunlight that gets fixed by net primary production:

$$\eta_{net} = \frac{Energy\ fixed\ by\ net\ primary\ production}{Energy\ in\ incident\ sunlight}$$

2.3.3.2 Measurement of primary production

The amount of primary production can be measured using different methods:

1. By measuring volumes of gases utilised and released:

This is done by examining the equation of photosynthesis

$$6CO_2 + 6H_2O \xrightarrow[\text{Solar energy}]{\text{Chlorophyll, enzymes}} C_6H_{12}O_6 + 6O_2$$

This can be put into energetics terms as:

$$6CO_2 + 6H_2O \xrightarrow[\text{2966 kJ}]{\text{Chlorophyll, enzymes}} C_6H_{12}O_6 + 6O_2$$

implying that for each mole of glucose produced,

- 2966 kJ of energy is absorbed
- 6 moles of CO_2 are utilised (134.4 litres at standard temperature and pressure[1])
- 6 moles of O_2 are released (134.4 litres at standard temperature and pressure)

These values can easily be measured to estimate primary productivity.

---

[1] Since 1982, STP is defined as a temperature of 273.15 K and an absolute pressure of exactly 100 kPa (1 bar).

2. By using radioactive decay of carbon:

The equation of photosynthesis is

$$6CO_2 + 6H_2O \xrightarrow[\text{Solar energy}]{\text{Chlorophyll, enzymes}} C_6H_{12}O_6 + 6O_2$$

In this process, we may replace CO_2 with labelled, radioactive $^{14}CO_2$.

After some time, the complete plant is harvested and the quantity of ^{14}C is measured to estimate the amount of CO_2 absorbed by the plant. Since absorption of carbon occurs through photosynthesis, the amount of carbon absorbed can give us an estimation of the amount of primary production done by the plant.

One shortcoming of this method is that some amount of ^{14}C may also get lost during respiration, so the method only measures the net primary production.

3. Harvest method:

The amount of plant material produced may be measured as:

$$\Delta B = B_2 - B_1$$

where

ΔB = change in biomass between times t_2 and t_1, an estimate of biomass produced (or lost), also an estimate of the amount of primary production done by the plants

B_2 = biomass at time t_2

B_1 = biomass at time t_1

4. By using satellite data and allometric equations:

We can also make estimates of primary productivity using satellite data by measuring chlorophyll and the level of different gases at different locations [Fig. 2.8a] and through computer simulations [Fig. 2.8b] by computing the production for given climatic conditions, water availability, site quality (fertility), efficiency and known allometric relations[2] for different species of plants.

2.3.3.3   What does primary productivity depend upon?

Primary productivity is a function of seven variables [Monteith, 1972]:

1. Solar constant: The rate at which energy reaches the Earth's surface from the Sun, usually taken to be 1,361–1,362 watts per square metre. If more energy reaches the Earth from the Sun, more will be the amount of primary production. The amount of energy given out by the Sun varies in a cyclical manner (the 11-year Solar cycle).

2. Latitude: Areas at higher latitudes (towards the poles) receive sunlight at slanted angles, reducing the effective amount of sunlight. Thus, the amount of primary productivity decreases as we move from the equator towards the poles.

3. Cloudiness: Clouds obstruct sunlight and reduce primary productivity.

---

[2]Allometry is the growth of body parts at different rates, resulting in a change of bodily proportions. In this case we are interested in how a plant distributes biomass into its body parts. If the biomass is utilised to make leaves, the productivity will increase at a much faster rate than if the biomass is utilised for lateral growth of the plant.

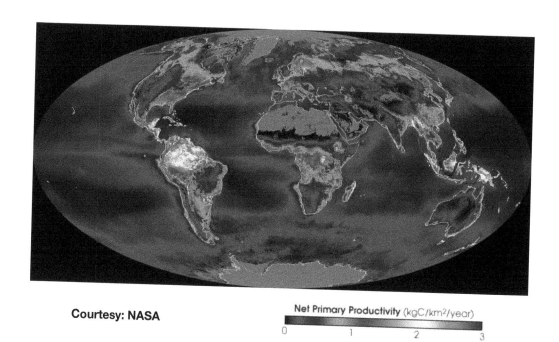

Courtesy: NASA

(a) Estimation of primary productivity through satellite data [NASA, 2003].

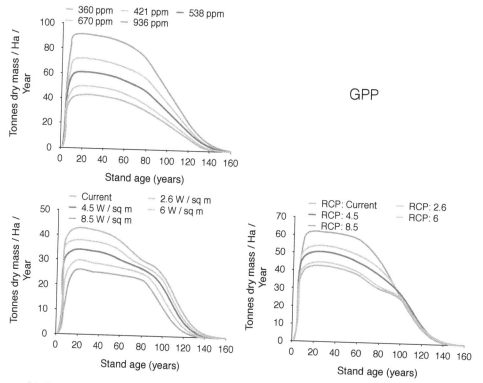

(b) Estimation of primary productivity through computer modelling [Awadhiya, 2017].

Figure 2.8: Estimation of primary productivity.

4. Dust and water in the atmosphere: These obstruct sunlight and reduce primary productivity.

5. Leaf arrangement: Plants with leaf arrangements that maximise the gathering of sunlight will have higher productivity than plants with sub-optimal leaf arrangements.

6. Leaf area: Plants with greater leaf area will have higher productivity than plants with lesser leaf area.

7. Concentration of CO_2 and other nutrients: Since CO_2 and other nutrients are raw materials for primary production, a scarcity of these will reduce primary productivity.

2.3.4 Nutrients

Nutrients are substances used by an organism to survive, grow and reproduce. We get nutrients such as carbohydrates, fats, proteins, fibre, etc. from the food that we eat. Primary producers get their nutrients from the environment, including

1. sediments in rivers

2. bird droppings

3. upwelling in oceans

4. dust plumes [Fig. 2.9], etc.

Figure 2.9: Dust plumes bringing sediments to the ocean [NASA, 2009].

These nutrients are fixed by primary producers into food, which then moves through food chains to reach other organisms, including us.

Since plants are dependent on the supply of nutrients from the environment, the amount of nutrients in an area has a large bearing on the number of plants that can be supported in that area, manifested as the plant density in that area. For example, a lake ecosystem — on the basis of the amount of nutrients and thus primary productivity — can be classified into

1. Oligotrophic lake: A lake with low primary productivity as a result of low nutrient content. These are characterised by low algal production, and often have very clear waters as represented by large Secchi depths,[3] with high drinking-water quality. These are generally found in high mountains. Good examples are glacial lakes that have nearly pure water derived from melting of snow and ice. This water is nearly free of any minerals or other nutrients.

2. Mesotrophic lake: A lake with an intermediate level of productivity. These lakes are often clear water lakes and ponds with beds of submerged aquatic plants and medium levels of nutrients.

3. Eutrophic lake: A lake with high biological productivity due to excessive concentrations of nutrients, especially nitrogen and phosphorus.

4. Hypereutrophic lake: An extremely nutrient-rich lake that can support a very high density of plants. It is characterised by frequent and severe nuisance algal blooms that arise to use the high nutrients, resulting in low transparency, high acidity and low oxygen levels when the algae die and get decomposed. The low oxygen levels form an anoxic environment that often results in creation of dead zones beneath the surface which kill all life forms in the lake. Examples include lakes with inflows from agricultural fields bringing in large doses of fertilisers.

The impacts of the amounts of nutrients can thus be seen in several water parameters, as shown in table 2.3.

Table 2.3: Impacts of nutrient levels.

Trophic class	Trophic Index	Chlorophyll (μg / litre)	Phosphorus (μg / litre)	Secchi depth (m)
Oligotrophic	$< 30 - 40$	$0 - 2.6$	$0 - 12$	$> 8 - 4$
Mesotrophic	$40 - 50$	$2.6 - 20$	$12 - 24$	$4 - 2$
Eutrophic	$50 - 70$	$20 - 56$	$24 - 96$	$2 - 0.5$
Hyper-eutrophic	$70 - 100+$	$56 - 155+$	$96 - 384+$	$0.5 - < 0.25$

2.3.4.1 Kinds of nutrients

On the basis of amounts needed, nutrients are classified into macronutrients and micronutrients:

1. Macronutrients: Nutrients needed in large amounts. These include

   (a) Macronutrients derived from air and water
       i. Carbon
       ii. Hydrogen
       iii. Oxygen
   (b) Primary macronutrients
       i. Nitrogen
       ii. Phosphorus

---

[3] Secchi depth is a measure of transparency or turbidity in a water body. The more the depth, the clearer is the water. The less the depth, the more turbid is the water in the water body.

iii. Potassium

(c) Secondary and tertiary macronutrients

    i. Sulphur

    ii. Calcium

    iii. Magnesium

2. Micronutrients or trace elements: Nutrients needed in small or trace amounts. These include

(a) Iron	(f) Sodium	(k) Aluminium
(b) Molybdenum	(g) Zinc	(l) Silicon
(c) Boron	(h) Nickel	(m) Vanadium
(d) Copper	(i) Chlorine	
(e) Manganese	(j) Cobalt	(n) Selenium

On the basis of their being obligatory (meaning that the organism cannot do without it) or non-obligatory, nutrients are classified into essential and non-essential nutrients. The criteria for an element to be classified as an essential element for plants are:

1. In the absence of the element the plants should be unable to complete their life cycle.

2. The deficiency of an essential element cannot be met by supplying some other element.

3. The element must be directly involved in the metabolism of the plant.

The roles of some prominent essential elements are

1. Nitrogen: constituent of proteins, nucleic acids, vitamins, hormones

2. Phosphorus: constituent of nucleic acids, ATP, cell membrane, certain proteins

3. Potassium: cation-anion balance needed for maintaining cell turgidity, opening and closing of stomata, activation of certain enzymes

4. Calcium: calcium pectate in cell wall, activation of certain enzymes, calcium channels in cell membranes

5. Magnesium: constituent of chlorophyll, activation of respiration enzymes

6. Sulphur: constituent of amino acids cysteine and methionine, several vitamins and coenzymes

These nutrients are made available to the plants from the environment, particularly land, air and water. However, over the billions of years of plant life on the Earth, these nutrients would have all gotten exhausted, since the Earth does not have an infinite supply of nutrients. The way nature is able to continue the supply of nutrients is through the biogeochemical cycles that maintain a continuous supply of nutrients to meet the needs of organisms.

2.3.5 Biogeochemical cycles

A biogeochemical cycle is a pathway by which a chemical substance moves through biotic (biosphere) and abiotic (lithosphere, atmosphere, and hydrosphere) compartments of the Earth [Fig. 2.10]. Biogeochemical cycles are also known as nutrient cycles, since they provide a continuous supply of nutrients to the biosphere.

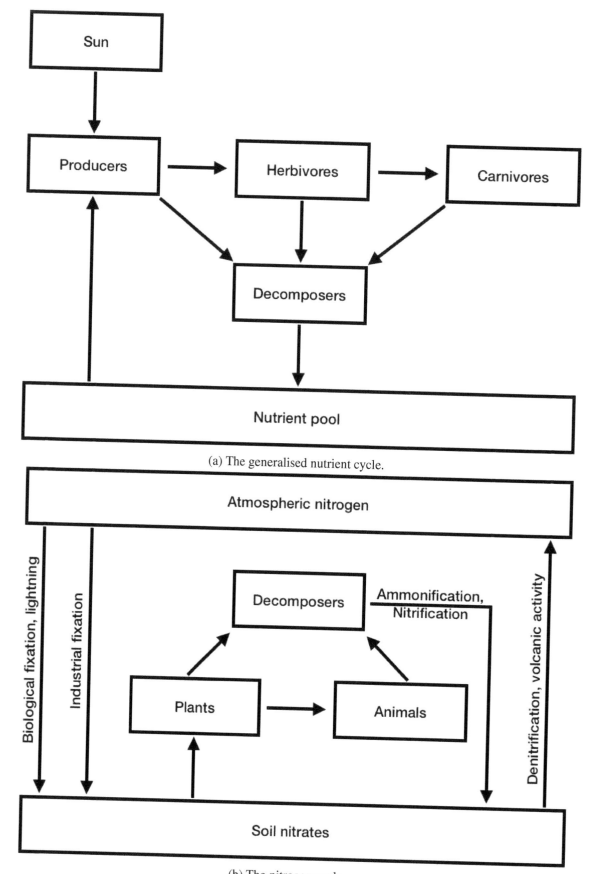

(a) The generalised nutrient cycle.

(b) The nitrogen cycle.

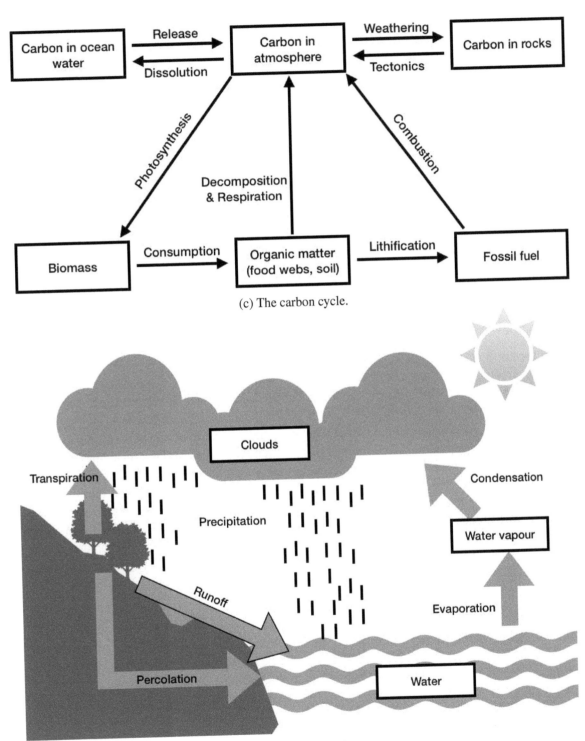

(c) The carbon cycle.

(d) The water cycle.

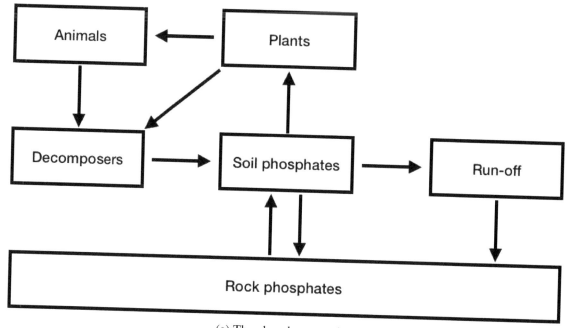

(e) The phosphorus cycle.

Figure 2.10: Examples of nutrient cycles.

2.3.5.1 The Generalised nutrient cycle

The generalised nutrient cycle is shown in figure 2.10a. The producers take energy from the Sun (or the environment in the case of chemoautotrophs) and elements from the nutrient pool to make food. The nutrient pool is the sum of all sources of nutrients that is available to the producers. These nutrients can be in the lithosphere (earth), hydrosphere (water) or the atmosphere (air). Once the producers have produced food, these nutrient elements are now incorporated in the food molecules. When herbivores eat plants or their parts, these nutrients become available to them. When the herbivores get eaten by carnivores, the nutrients become available to the carnivores. Thus, the nutrients are able to move from the abiotic nutrient pool to the biosphere. When the producers, herbivores or carnivores die, or shed their body parts such leaves or skin, or defecate, the nutrients present in those biotic parts are acted upon by the decomposers. In this process, the decomposers get their food — nutrients and energy. The decomposers finally break down the food and release the elements back into the nutrient pool. And the cycle continues as the nutrients are then again taken up by the producers to make food.

2.3.5.2 The Nitrogen cycle

The nitrogen cycle is shown in figure 2.10b. There are two nutrient pools of nitrogen: nitrogen in the atmosphere and soil nitrates. Nitrogen moves from the atmospheric pool to the soil pool through nitrogen fixation — the conversion of atmospheric nitrogen into ammonia or related nitrogenous compounds, which occurs by

1. Biological fixation:

$$N \equiv N \xrightarrow{\text{Nitrogenase}} NH_3$$

which is done by organisms such as

(a) *Rhizobium*: symbiotic bacteria

(b) *Azotobacter*: free-living bacteria

(c) *Nostoc*: cyanobacteria

(d) *Anabaena*: cyanobacteria

The ammonia is then biologically oxidised to nitrites and nitrates in a process called nitrification:

$$2NH_3 + 3O_2 \xrightarrow{\text{Nitrosomonas or Nitrococcus}} 2NO_2^- + 2H^+ + 2H_2O$$

$$2NO_2^- + O_2 \xrightarrow{\text{Nitrobacter}} 2NO_3^-$$

done by chemoautotrophic nitrifying bacteria.

2. Lightning [Ferguson and Libby, 1971]: Lightning generates ions and converts atmospheric nitrogen into nitrous acid which falls to the Earth along with rains. The mechanism consists of three stages:

(a) Production of NO^+

$$O^+ + N_2 \rightarrow NO^+ + N$$

$$N_2^+ + O \rightarrow NO^+ + N$$

$$N^+ + O_2 \rightarrow NO^+ + O$$

$$O_2^+ + N \rightarrow NO^+ + O$$

$$O_2^+ + N_2 \rightarrow NO^+ + NO$$

$$(N^+, N_2^+, O^+, O_2^+, NO_2^+) + NO \rightarrow (N, N_2, O, O_2, NO_2) + NO^+$$

(b) Hydration of NO^+ to form $NO^+(H_2O)_3$

$$NO^+ + H_2O + M \rightarrow NO^+H_2O + M$$

$$NO^+H_2O + H_2O + M \rightarrow NO^+(H_2O)_2 + M$$

$$NO^+(H_2O)_2 + H_2O + M \rightarrow NO^+(H_2O)_3 + M$$

(c) Formation of HNO_2

$$NO^+(H_2O)_3 + H_2O \rightarrow H_3O^+(H_2O)_2 + HNO_2$$

The HNO_2 then falls down to the Earth with rains.

3. Industrial fixation: Two processes are generally used to fix nitrogen industrially, especially for use as fertilisers:

(a) Haber process

$$N_2 + 3H_2 \xrightarrow[\text{High temperature and pressure}]{\text{Catalyst}} 2NH_3$$

(b) Ostwald process

$$4NH_3 + 5O_2 \xrightarrow{\text{Catalyst}} 4NO + 6H_2O$$

$$2NO + O_2 \xrightarrow{\text{Catalyst}} 2NO_2$$

$$4NO_2 + O_2 + 2H_2O \xrightarrow{\text{Catalyst}} 4HNO_3$$

Once the nitrogen is 'fixed,' it is available for use by plants. A part of the 'fixed' nitrogen in the form of nitrates may also remain locked up in the minerals found in rocks. This may be slowly released back through weathering (or even through volcanic activity). In the biotic world, nitrogen is a constituent of proteins, nucleic acids, vitamins and hormones. When animals eat plants, nitrogen reaches the animal kingdom, and then moves from primary consumers to secondary consumers, tertiary consumers, and so on till it reaches the apex predator. When plants and animals die, or shed body parts, or defecate, the nutrients present in those biotic parts are acted upon by the decomposers which use these biotic parts as their food. In this process, the decomposers finally break down the compounds and release the nitrogen back into the nutrient pool:

$$\text{Dead plants and animals} \xrightarrow{\text{Ammonification}} NH_3$$

The production of ammonia through the decomposition of organic nitrogen in dead plants and animals is called ammonification. This ammonia is then further converted into nitrogen — and released into the atmospheric pool — through the process of oxidation, or may be converted back into nitrites and nitrates in the soil pool.

The nitrogen in the soil pool may also get released into the atmospheric pool through the process of denitrification and through volcanic activity, completing the cycle.

2.3.5.3 The Carbon cycle

The carbon cycle is shown in figure 2.10c. There are four nutrient pools of carbon: carbon in ocean water, carbon in atmosphere, carbon in rocks (as carbonates) and carbon in fossil fuel. Carbon moves through all these pools and through the biosphere.

Carbon in the atmosphere can dissolve in water. When water heats up, this carbon gets released back into the atmosphere. Similarly, carbon in the atmosphere can get trapped into rocks through the process of weathering:

$$CaO + CO_2 \xrightarrow{\text{Carbonation}} CaCO_3$$

$$CaCO_3 + CO_2 + H_2O \xrightarrow{\text{Weathering}} Ca(HCO_3)_2$$

Similarly, carbon in the rocks can get released when heated due to tectonic processes and volcanic activity:

$$CaCO_3 \xrightarrow{\text{Heating}} CaO + CO_2$$

Carbon in the atmosphere — present as carbon dioxide — enters the biosphere through the process of photosynthesis by photoautotrophs, followed by their consumption by several organisms in many food chains. All these organisms release carbon dioxide during the process of respiration, and when their bodies are decomposed after death. Through these processes, carbon moves from the biosphere back into the atmosphere.

Some carbon in the biosphere also gets converted into rocks through the process of lithification to form fossil fuels such as coal, petroleum and natural gas. When these fossil fuels are extracted and burnt, the stored carbon gets released back into the atmosphere as carbon dioxide, completing the cycle.

2.3.5.4 The Water cycle

The water cycle is shown in figure 2.10d. There are three pools of water: water in water bodies (hydrosphere), water in atmosphere (as water vapour) and water in lithosphere (as ground water). A small portion of water also remains as chemically incorporated water in several compounds.

Water from the water bodies gets evaporated due to heat of the Sun, rises, and condenses upon cooling to form clouds. These clouds can move to other areas with the wind. Through the process of precipitation, the water in the clouds falls down as rainwater. This may directly fall over water bodies, or over the land. The rain falling over land may reach back to the water bodies through runoff, or get absorbed in the soil in a process called percolation. The percolated water may move to the water bodies due to gravity, or may be absorbed by plants. A part of this water is used by the plants in the process of photosynthesis, while most of it is released back into the atmosphere through the process of transpiration. Animals get water through plants, or directly by drinking. This water is released back during breathing, perspiration, micturition and death.

2.3.5.5 The Phosphorus cycle

The phosphorus cycle is shown in figure 2.10e. The major nutrient pool is rock phosphates. Weathering releases the phosphates from the rocks into the soil, forming soil phosphates. Precipitation of soil phosphates converts them back into rock phosphates.

From the soil, the phosphates are taken up by plants and reach the biosphere. They move through animals and decomposers which release them back into the soil as soil phosphates. Some portion of the phosphates in the soil gets run-off into water bodies and, upon precipitation, forms rock phosphates, completing the cycle.

2.4 INTERACTIONS

The different organisms that live together interact with each other. Interactions are effects that the organisms in a community have on each other. They are of several types [Table 2.4]:

1. Intraspecific vs. interspecific interactions

   (a) Intraspecific interactions: Effects that the organisms in a community have on members of their own species.
   (b) Interspecific interactions: Effects that the organisms in a community have on members of species other than their own.

2. Harmonious vs. inharmonious interactions

   (a) Harmonious interactions: Positive ecological interactions where none of the participating organisms is harmed.
   (b) Inharmonious interactions: Negative ecological interactions where at least one of the participating organisms is harmed.

Table 2.4: Kinds of interactions.

	Harmonious	**Inharmonious**
Intraspecific	Intraspecific harmonious	Intraspecific inharmonious
Interspecific	Interspecific harmonious	Interspecific inharmonious

2.4.1 Intraspecific harmonious interactions

These are effects that the organisms in a community have on members of their own species in such a manner that none of the participating organisms is harmed.

Main intraspecific harmonious interactions are

1. Colonies: Colonies are defined as "functional integrated aggregates formed by individuals of the same species." Examples include coral reefs [Fig. 2.11a], filamentous algae and microbial colonies [Fig. 2.11b]. Colonies being large structures are much more formidable than individual organisms, and so provide a large amount of protection to all the members in the colony. Also, through combined action of several organisms, colonies are much more efficient than individual organisms in obtaining food, thus benefitting all the members of the colony. Since colonies are formed by individuals of the same species, and the individuals receive the benefits of protection and nutrition, colonies are good examples of intraspecific harmonious interactions.

2. Societies: Societies are defined as "interactions for labor division and collaboration among individuals of the same species." Examples include bee hives, termite mounds [Fig. 2.11c], ants [Fig. 2.11d] and wolf packs. By division of labour and collaboration, the efficiency of the whole society increases. Thus, in a bee hive, certain bees can specialise in gathering food, others in protecting the hive, yet others in maintaining the hive and the young ones, and yet others in reproduction. This specialisation increases efficiency, and the fruits of labour are shared to everyone's benefit. When worker bees gather food efficiently, more food becomes available to all the members of the bee hive. When the bee hive is protected, all the members get protection, and so on. Similarly, in a wolf pack, the individuals cooperate and communicate with each other to vastly increase the hunting efficiency of the whole pack. Thus, since societies are formed by individuals of the same species, and the individuals receive the benefits of protection, nutrition, etc., societies become good examples of intraspecific harmonious interactions.

2.4.2 Interspecific harmonious interactions

These are effects that the organisms in a community have on members of some other species in such a manner that none of the participating organisms is harmed.

Main interspecific harmonious interactions are

1. Protocooperation: Protocooperation is "an ecological interaction in which both participants benefit but which is not obligatory for their survival." Examples include birds eating ectoparasites on the bodies of animals [Fig. 2.11e] where the birds get food and the animals get rid of ectoparasites, cleaner fishes getting their food from the mouth of other organisms while providing them with a cleaning facility, hermit crabs and sea anemones living together with hermit crabs providing transportation to sea anemone and sea anemone providing protection to hermit crab, and pollination by insects [Fig. 2.11f] and birds [Fig. 2.11g] with flowers providing food to, and receiving benefits of transport of pollens from insects and birds. The

interaction is not obligatory for survival, meaning that even if the organisms involved in the interaction did not interact, they shall survive, albeit with lesser efficiency or benefits.

2. Mutualism: Mutualism is "an ecological interaction in which both participants benefit and which is obligatory for their survival." Examples include microbes digesting cellulose in the stomach of ruminants and *Rhizobium* in the root nodules of leguminous plants. In these cases, the microbes provide food to the host, and the host provides shelter and food to the microbes. Thus, members of two species derive benefit from the interaction. The interaction is obligate, meaning necessary for the survival of both the species. If they do not live together, they will not survive. For example, if the cellulose-digesting microbes are removed, the ruminants will be unable to digest their food, and die of starvation. Similarly, without the protection and food provided by the ruminants, the cellulose-digesting microbes will die.

3. Commensalism: Commensalism is "an ecological interaction in which one individual benefits while the other is neither benefits nor is harmed." Examples include bacteria and other micro organisms living on the skin without being pathogenic or beneficial, egrets feeding with buffaloes [Fig. 2.11h] and other herbivores [Fig. 2.11i,j] and birds hitching a ride on top of other animals [Fig. 2.11k]. Thus, while the grazing activity of buffaloes creates movement and brings out insects that the egrets can eat, providing a benefit to the egrets, the presence of the egrets does not provide any benefit or harm to the buffaloes. Thus, this is an example of commensalism.

2.4.3 Intraspecific inharmonious interactions

These are effects that the organisms in a community have on members of their own species in such a manner that at least one of the participating organisms is harmed.

Main intraspecific inharmonious interactions are intraspecific competition and cannibalism.

(a) An example of colony: coral.

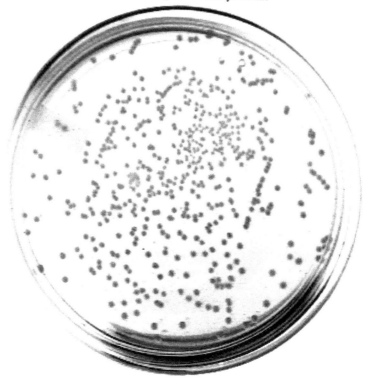

(b) An example of colony: bacterial colonies on a petri dish.

(c) An example of society: termite mound.

(d) An example of society: ants.

(e) An example of protocooperation: birds eating ectoparasites on giraffe's body in Kruger National Park.

(f) An example of protocooperation: pollination by insect in Kirstenbosch Botanical Garden, Cape Town.

(g) An example of protocooperation: pollination by a hummingbird in Kirstenbosch Botanical Garden, Cape Town.

(h) An example of commensalism: egrets with buffaloes.

(i) An example of commensalism: egrets with wild water buffaloes in Manas Tiger Reserve.

(j) An example of commensalism: egrets with rhinoceros in Manas Tiger Reserve.

(k) An example of commensalism: bird hitching a ride on a Cape Buffalo in Kruger National Park.

Figure 2.11: Examples of harmonious interactions.

Competition

Competition is "the ecological interaction in which individuals explore the same ecological niche or their ecological niches partially coincide and, therefore, competition — or fight — for the same environmental resources takes place." Competition is of several types:

1. Intraspecific (between individuals of the same species) vs. interspecific (between individuals of different species).

2. Exploitative vs. interference [Table 2.5].

3. Apparent: an interaction between two prey species with a common predator. An increase in the population of one prey species may lead to an increase in the abundance of the common predator, leading to a stronger predation pressure on the second prey species. In this manner, the two prey species have a relation of indirect competition between them, mediated by the numerical response of the common predator species.

Such examples are observed in

    (a) insect host-parasitoid communities [Holt and Lawton, 1993].

    (b) exotic shrubs and trees through the action of seed predators [Meiners, 2007].

    (c) grasses and plants through rodents [Dangremond et al., 2010].

Thus, when grasses produce seeds in large numbers, the rodent population grows, which then feeds upon the seeds of other plants as well, harming them more due to their numerical abundance. Apparent competition results in the reduction of the prey species' equilibrium densities and growth rates, and is a commonly observed in numerous food webs [Holt and Bonsall, 2017].

Table 2.5: Differences between exploitative and interference competition.

Exploitative competition	Interference competition
Also called "scramble competition."	Also called "contest competition."
Competition is exploitative when species or individuals compete for the same limited resource.	Competition is interference when species or individuals deplete each other's resources by interferences such as aggressive displays or fighting.
In exploitation, organisms use up resources directly, so it is not longer available for use by others.	In interference, one organism prevents others from using the resources.
There is no direct contact or conflict between the species or individuals in exploitation.	There is direct contact or conflict between the species or individuals in interference.
Competitive ability in exploitation is the rate of resource consumption.	Competitive ability in interference is the ability to put on aggression or fights.
Pure exploitative competition can be modelled as affecting the carrying capacity.	Pure interference competition can be modelled as affecting the rate of increase per individual.
For pure exploitative competition the relation between the rate of change per individual of one species and abundance of the second is non-linear.	For pure interference competition the relation between the rate of change per individual of one species and abundance of the second is linear.
e.g. 1. intraspecific: an organism overgrazing on a land shared by several individuals of the species. 2. interspecific: canopy trees of several species competing for the available sunlight.	e.g. 1. intraspecific: an animal showing territorial behaviour to its conspecifics. 2. interspecific: allelopathy.

Cannibalism is "the act of one individual of a species consuming all or part of another individual of the same species as food." Examples include sexual cannibalism in black widow, sexual cannibalism in praying mantis and eating of tiger cubs by other (incoming) male tigers to make the mother tiger receptive to mating.

2.4.4 Interspecific inharmonious interactions

These are effects that the organisms in a community have on members of some other species in such a manner that at least one of the participating organisms is harmed.

Main interspecific inharmonious interactions are interspecific competition [Fig. 2.12a], parasitism, predation and amensalism. Interspecific competition is already described in subsection 2.4.3.

Parasitism is "an ecological interaction in which an organism lives at the expense of another." Parasites can be ectoparasites — living outside the body of the host — such as leech, or endoparasites — living inside the body of the host — such as *Plasmodium vivax*. Parasitism is widely observed in both plants and animals.

Predation is "an ecological interaction in which one individual mutilates or kills another to get food." Examples include several birds [Fig. 2.12b,c], predators like lions and tigers, and insectivorous plants [Fig. 2.12d].

Amensalism is "an interaction where an organism inflicts harm to another organism without any costs or benefits received by itself." A good example is the trampling of grass due to movement of animals [Fig. 2.12e] which harms the grass, but neither harms nor benefits the animal.

(a) Invasive species such as *Lantana camara* are often more competitive than native vegetation.

(b) An example of predation: bird feeding on fish in Manas Tiger Reserve.

(c) An example of predation: a roller bird feeding on centipede in Kruger National Park.

(d) Pitcher plants are insectivorous plants and can widely be seen in Mukurthi National Park.

(e) An example of amensalism in Blackbuck National Park, Velavadar.

Figure 2.12: Examples of inharmonious interactions.

CHAPTER 3

Population growth and community organisation

The individuals of a species that live together in a place form a population. Thus, Kanha Tiger Reserve has a population of tigers. If we wish to conserve tigers, we should know about how the population is doing — is it increasing, decreasing or remaining constant? And what are the projections of this population in the near and distant future? How should we manage this population in a manner that the population grows with time, or at least remains consistent? Questions such as these form a part of the discipline of Population Ecology.

At the same time, it is important to appreciate that the tiger population is not the only population in the area. Kanha has other predators — like leopards and wolves. These predators may feed on the same prey that the tiger feeds upon. In an extreme situation, it is possible that they may result in a food shortage of tigers. Similarly, Kanha also has a large number of prey populations — like those of sambar, deer and the Indian bison. These populations have their own dynamics. And the size of these populations will determine whether the tigers have enough to feed on. Hence, when we wish to work on tiger conservation, we need to manage not only the tiger population, but also the populations of several other species. This is because tigers interact with, and dependent upon, and are affected by other species in the area. These populations living together in an area, and interacting together form a community. And the study of communities comes under the scope of Community Ecology.

In this lecture, we shall explore topics in Population and Community Ecology.

3.1 POPULATION AND CHANGES IN POPULATION

Population is defined as "all the organisms of the same group or species, which live in a particular geographical area, and have the capability of interbreeding." Thus when many individuals of the same species occupy the same geographical area, and have the capability of interbreeding, they become a population [Fig. 3.1].

Population is dynamic — it changes as some individuals are born, some die, and some move in or out. We can express it in the form of an equation as:

$$P_{n+1} = P_n + \text{Births} + \text{Immigration} - \text{Deaths} - \text{Emigration}$$

where
P_{n+1} = population at time $n+1$
P_n = population at time n
Births = number of individuals born into the population between time n and $n+1$
Deaths = number of individuals in the population that die between time n and $n+1$

(a) An individual wild water buffalo in Kaziranga Tiger Reserve. The individual forms a unit of the population.

(b) A wild water buffalo population in Kaziranga Tiger Reserve. This group lives together in a particular geographical area, and has the capability of interbreeding.

(c) A flamingo population in Wild Ass Sanctuary.

Figure 3.1: What is a population?

Immigration = number of individuals that move into and become part of the population between time n and $n+1$

Emigration = number of individuals that move out of and cease to remain part of the population between time n and $n+1$

A major activity in conserving wildlife is to determine how a population is performing over time. This assessment of wildlife populations is important for several reasons:

1. Numbers are essential at every stage of management. We may express management interventions through the Deming cycle:

   Plan → Do → Check → Act

   Numbers are critically required at all of these stages, to understand whether or not the management intervention is working as it should.

2. Numbers are crucial inputs for decision support. Depending on existing numbers of wildlife, we may need to:

   (a) increase their numbers if the numbers are getting low, or

   (b) reduce their numbers if the numbers are very high — say over the capacity of the environment, or when there are situations of conflict, or

   (c) maintain the status-quo if the numbers are adequate

   Which of these is to be followed can best be determined by an assessment of wildlife numbers and their impacts.

3. Numbers help assess the risk of a population decline or crash. When the number of individuals is very low, the population might crash. Reducing numbers can forewarn us about such an impeding decline.

4. Numbers help us plan scenarios and take steps to counter any unwanted changes. The steps could be:

   (a) adaptation, where the population is made supple enough to respond to changes, or

   (b) mitigation, where the causes of change are analysed and addressed

3.1.1 Do we need numbers or trends?

Actually, we need both.

Trends are helpful when we need to analyse and address the gross movement of population numbers — whether the population is increasing, decreasing or remaining constant. This is especially important for, say, prey species like chital or sambar, where exact numbers are hard to compute due to their large population sizes.

On the other hand, **numbers are of crucial importance in cases where the populations have reached critical limits, or when the animal is a priority species.** Thus, we need to have census for tigers, lions, Great Indian bustards and dugongs. Mere trends will not suffice.

It is also important to note that numbers are finer data than trends. With exact numbers, we can easily compute trends. The information on trends alone is not sufficient to deduce the exact number of individuals. But this comes at a cost — in terms of time, effort and money required to get the exact numbers. When we have a shortage of resources, trends become a 'quick and dirty' method to get a grasp of wildlife populations.

3.1.2 What demographic information are we interested in?

As managers and scientists, we are interested in several demographic parameters describing the population, such as:

1. **population density**: the number of individuals in the population per unit area (generally per hectare or per square kilometre)

2. **population size**: the number of individuals in the population

3. **age pyramid**: the distribution of various age groups in the population

4. **crude birth rate**: the annual number of live births per 1,000 individuals

5. **crude death rate**: the annual number of deaths per 1,000 individuals

6. **general fertility rate**: the annual number of live births per 1,000 females of reproductive age

7. **age-specific fertility rate**: the annual number of live births per 1,000 females of specific age classes in the reproductive age

8. **total fertility rate**: the average number of live births per female individual completing her reproductive age, if she followed the current age-specific fertility rate of the population

9. **replacement level fertility**: the average number of offsprings that a female individual must produce such that the population is completely replaced for the next generation. The replacement level of fertility ≥ 2

10. **juvenile mortality rate**: the annual number of deaths of juveniles per 1,000 live births

11. **life expectancy**: the number of years that an average individual in the population at a given age could expect to live, at the present age-specific mortality levels

3.1.3 Population density

Of the several characteristics of populations, population density is one of the most important. It is generally easier to compute than a total count of all the individuals in an area, and gives a good indication of the population trends. Population density is defined as:

$$\text{Population density} = \frac{\text{Number of individuals in an area}}{\text{Area}}$$

Thus, number of individuals in an area (population count) can be computed from population density as

$$\text{Number of individuals in an area} = \text{Population density} \times \text{Area}$$

We define two kinds of population density

1. Absolute density: Number of individuals per unit area, given as number per sq. km or number per hectare.

2. Relative density: Whether area x has more organisms than area y, or vice-versa? Here we only do a comparison of population density in two areas, without actually measuring the absolute numbers in any area.

Measurement of absolute population density can be done through

1. Total counts, e.g. census. In this method, the area is cordoned off and all the individuals in the area are counted. Then, population density is determined as

$$\text{Population density} = \frac{\text{Number of individuals in the area}}{\text{Area}}$$

This method is generally used only for very small populations that can easily be counted with a high level of accuracy. Since the method is labour, cost and time-intensive, it is only used for critical species, or for academic purposes.

2. Sampling methods: In these methods, a small, representative portion of the population is assessed for density measurements, and the values obtained are considered to represent the complete population. Sampling methods include

    (a) Quadrats: "each of a number of small areas of habitat selected at random to act as samples for assessing the local distribution of plants or animals," which can be

        i. square
        ii. rectangular
        iii. circular
        iv. irregular

    (b) Capture-recapture method: In this method, some individuals from the population are captured at time t_1, marked — say with tags, paints, or using body marks, and released back into the population. They are allowed to freely mix with the population. At a later time t_2, some individuals from the population are captured again, and checked for markings that were put at time t_1.
    Only a fraction of the individuals captured at time t_2 will have the markings. If the animals had mixed freely, we can compute the total number of individuals as:

$$\frac{\text{No. of marked animals in sample}}{\text{No. of animals caught in sample}} = \frac{\text{No. of marked animals in population}}{\text{Total population size}}$$

This method is based upon several assumptions:

i. Marked and unmarked animals are captured randomly and without preference. Thus, the animals that get captured should not be different from the animals that were not captured. This assumption gets violated when some animals have behaviours that make them easy to capture (also called as trap-happy animals, such as those with more inquisitiveness or with daring nature), while some others have behaviours that make them difficult to capture (also called as trap-shy animals, such as those that are timid and uninterested).

ii. Mortality rate in marked and unmarked animals is the same: This assumption may get violated due to rough handling, conditions such as capture myopathy [Section 7.6] or if the animals are given vitamins or antibiotics before release.

iii. Marks are not lost or overlooked: This assumption gets violated when tags get lost, or when the body markings or colour are not visible due to mud or shade.

Several species have natural, individual specific marks that may be used as permanent marks in capture-recapture technique. Some such species are:

i. Striped hyena, *Hyaena hyaena*: using the patterns of stripes on the body

ii. Leopards: common (*Panthera pardus fusca*), clouded (*Neofelis nebulosa*), snow (*Panthera uncia*): using the patterns of rosettes on the body

iii. Tiger, *Panthera tigris*: using the patterns of stripes on the body

iv. Lion, *Panthera leo persica*: using their vibrissae patterns in the three rows of whiskers; less than 1 in 10,000 chance of two lions have the same pattern

v. Toads and frogs e.g. *Amolops formosus*: using the patterns of warts and body markings

vi. Cats, including leopard cat (*Prionailurus bengalensis*) and marbled cat (*Pardofelis marmorata*): using the body markings

vii. Spotted deer, *Axis axis*: using spots on the body

viii. Some marine mammals like humpback whales, *Megaptera novaeangliae*: using natural markings on the body

ix. Crocodiles such as *Crocodylus palustris*: using numbers and patterns of scutes

x. Snakes such as the python, *Python molurus molurus* Linn.: using the patterns of the first four blotches

3. Removal method: In this method, a portion of the area is cordoned off, and animals are caught and removed. With each passing day, with the same amount of effort, the number of animals removed will go on decreasing, since there are lesser and lesser number of animals available for removal. The total number of animals removed will go on increasing, but the cumulative curve will flatten out [Fig. 3.2]. The flat portion of the curve will give an estimate of the total number of animals present in the cordoned-off area, which can be used to compute population density in the area.

Mathematically, if the rate of decrease in population due to removal is given as

$$\frac{dN}{dt} = -aN$$

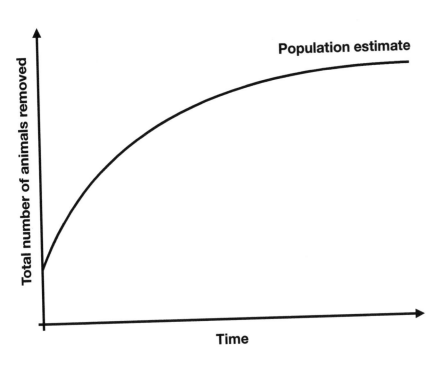

Figure 3.2: An illustration of the removal method.

then we have

$$N = N_0 e^{-at}$$

a and N_0 can be estimated using regression.

Measurement of relative population density can be done through a comparison of wildlife signs in the two areas, such as

1. number of faecal pellets in the two areas found per unit of effort

2. frequency of vocalisation of organisms observed in the two areas

3. pelt counts — number of animals captured by trappers in the two areas

4. catch per unit fishing effort in the two areas

5. number of artefacts, e.g. nests in the two areas

6. questionnaires to field staff in the two areas

7. percentage of plant cover in the two areas — more suitable is the habitat, more will be the animal density

8. frequency of quadrats where the species is found in the two areas (occupancy survey)

9. feeding capacity — the amount of bait taken by animals in the two areas, etc.

3.2 COMMUNITY ORGANISATION AND CHANGES

A community is "an assemblage of populations of living organisms in a prescribed area or habitat." Thus, if we have a population of hog deer [Fig. 3.3a] and another population of wild water buffalo [Fig. 3.3b] and they live together, they will form a community [Fig. 3.3c], possibly with several other populations as well.

A transition area where two communities meet and integrate together is known as an *ecotone*, which may be sharp [Fig. 3.4a] or diffuse [Fig. 3.4b].

A community has the following attributes:

1. Co-occurrence of species in the same area: A community of tigers and chitals is formed when tigers and chitals occur in the same area.

2. Recurrence of groups of the same species in different areas: Since tigers and chitals form a community in one area, there can be other areas — such as other tiger reserves — which will also have the same two species — tigers and chitals — forming a community.

3. Homeostasis or self-regulation: A community can often self-regulate. Thus, if the population of chitals goes down, the tigers will get less prey and the tiger population will reduce. When the tiger population goes down, the chitals will have less number of predators, and their population will bounce back. When the chital population increases, the tigers will get more food easily and their population will increase. When the tiger population increases, more number of chitals will get hunted and eaten and thus their population will go down.

We can observe in this case that the tigers and chitals in the community are able to regulate each other's numbers — whatever be the starting point. In other words — the community of tigers and chitals is able to self-regulate in such a manner that we neither get a population explosion, nor a population collapse, but only a continuity of the populations that form the community. From a management perspective, this explains why we should strive to have healthy communities — they ensure by themselves that the individual populations are taken care of — automatically and without the need for a manual intervention. We further elaborate this phenomenon in Section 3.3.3.

A community can be characterised by means of:

1. Species diversity: Which species are living together?

2. Growth form and structure of the community: Is it a grassland, a woodland, or a mixture in terms of trees, shrubs, herbs, mosses, etc.? What is the vertical stratification in the community?

3. Dominance: Dominant species are those that are "highly successful ecologically and which determine to a considerable extent the conditions under which the associated species must grow." A community may be characterised in terms of the dominant species — such as a sal-dominated community — which will be dominated by sal trees and have a specific set of species that can survive with sal.

4. Relative abundance: The relative proportions of different species in the community.

5. Trophic structure: Who eats whom in the community?

3.2.1 Community theories

There are three main theories of communities, moving from a rigid structure with sharp boundaries, through a random collection of organisms to organisation of species along a continuum:

(a) Population 1: Hog deer in Kaziranga Tiger Reserve.

(b) Population 2: Wild water buffalo in Kaziranga Tiger Reserve.

(c) A community in Kaziranga Tiger Reserve: assemblage of populations of living organisms in a prescribed area or habitat.

Figure 3.3: Population and community.

(a) A sharp ecotone: land-water boundary in Cape Town.

(b) A diffuse ecotone: land-water boundary in Nalsarovar Bird Sanctuary.

Figure 3.4: Sharp and diffuse ecotones.

1. Clements' organismic theory: It states that the community is a super-organism or discrete unit, with sharp boundaries. Communities exhibit properties which are more than the sum of the individual parts. Species interactions are assumed to play a major role in the development and regulation of communities in this theory.

2. Hubbell's neutral theory: It states that species are functionally equivalent, and the abundance of a population of a species changes by random births and deaths.

3. Gleason's individualistic /continuum theory: It states that the abundance of a population of a species changes gradually along complex environmental gradients. According to this view communities are not tightly structured, but are merely coincidences resulting from chance dispersal, environmental sorting and species interactions. Communities are less predictable, and species interactions have a much reduced role in determining the structure of communities.

Different communities may fit one of these models, and can then be characterised and understood through these theories. It is important to note here that given the variety in communities, we cannot have a 'one size fits all' approach to understanding community organisation. But we may group similar communities together to understand the fundamentals of their functioning. If we consider two communities, one with a number of species, and another with b number of species; and if there are c species that occur in both the communities, we state that the index of similarity between these communities is represented as:

$$\text{Index of similarity} = \frac{2c}{a+b}$$

Among the species that constitute the communities, it is observed that while some species are found in many communities, others are confined to a few communities — or even a single community. Based on their association with a community, species may be classified into [Blanquet and Pflanzensoziologie, 1964]:

1. Accidental species: Rare species in the community, present either as chance invaders from another community or as relicts from a previous community.

2. Indifferent species: Species with no real affinity for any particular community, but which are not rare.

3. Preferential species: Species that are present with varying abundance in several communities, but are especially abundant and vigorous in one particular community.

4. Selective species: Species found most frequently in a particular community, but also present occasionally in others.

5. Exclusive species: Species that are confined completely or almost completely to a particular community. These species often have an obligatory relationship with another species of the community.

3.2.2 Community stability

When a community is faced with changes — say in temperature or rainfall or due to a new disease — it responds to those changes. These responses can be classified into

1. Resistance — a measure of how little the variable of interest changes in response to external pressures. If the changes are small, we say that the community is able to *resist* those changes without any perceptible alterations in the community. This ability of the living system to resist external fluctuations is also known as inertia or persistence.

2. Resilience — If the changes are large and the community is unable to resist those changes, it will be altered. However, it may happen that after some time, the changes revert back (as in when it rains after a severe drought). Then the community may come back to its original form, or it may remain in its altered form. The tendency of a system to retain its functional and organisational structure after a perturbation or disturbance is known as resilience.

Thus, if a community is resilient, it will bounce back to its original state after the changes are reversed. Resilience is measured in terms of elasticity and amplitude.

    (a) Elasticity: It is the speed with which a system returns to its original / previous state. If the community is elastic, it will quickly return to its original state. If the community is inelastic, it will return back slowly.

    (b) Amplitude: It is a measure of how far a system can be moved from the previous state and still return. If a community has a large amplitude of resilience, it can return back after large changes. If a community has a small amplitude of resilience, it can only return back if the changes are small.

3. Collapse — An ecosystem is considered collapsed when its unique biotic (characteristic biota) or abiotic features are lost from their previous occurrences. In such cases, the community changes, and the changes are permanent. It may also result in a total annihilation or extermination of the community.

Community stability is the ability of a community to defy change (resistance) or to rebound from change (resilience). There is no direct relationship between diversity and stability, but several experiments have suggested that very simple systems are inherently unstable, and that diversity increases stability in several communities. This is often because in biodiversity-rich communities, even if one or a few species are removed, there are certain species available to take their place, and the community continues to exist.

3.2.3 Community evolution: Ecological succession

From the perspective of conservation, we may want communities to continue as they are, without any changes. However, not all changes to communities are undesirable. Some changes may make the communities more resilient and better able to utilise their surroundings. One such change is ecological succession that occurs naturally in several communities.

Ecological succession is defined as "the process of change in the species structure of an ecological community over time." In this process, an ecosystem changes from one that supports minimal (or no) life forms (also known as pioneer community) to one that is stable and supports a large biodiversity (also known as climax community). The intermediate stages are known as seral communities or seres, which can be of several kinds:

1. Hydrosere (Greek *hydro* = water) — a community in water

2. Xerosere (Greek *xero* = dry) — a community in dry areas, including

    (a) Lithosere (Greek *lithos* = rock) — a community on rock

    (b) Psammosere (Greek *psammos* = sand) — a community on sand

3. Halosere (Greek *halos* = salt) — a community in a saline body

Let us explore ecological succession by looking at an example. We begin with a bare rock [Fig. 3.5a]. This rock may have just formed out of lava, or may have arisen because a large rock has fractured, revealing a fresh surface. There are no life forms on this rock. In the absence of soil, it does not support plants. And in the absence of producers, the consumer animals can't be supported.

This rock — which lies bare — is however exposed to the natural elements. The Sun heats it up during the day, resulting in an expansion of the surface layer. During the night, the surface cools down and contracts. This repeated expansion and contraction results in several small cracks in this rock. The rock is breaking apart.

Rocks are made of minerals. When the rock breaks, these become available for use by life forms. Since we still do not have soil, we won't find plants. But simpler life forms — such as lichens — can be supported. They will form the pioneer community, for they are the first ('pioneers') to colonise the rocks. Lichens are composite organisms in which algae, bacteria and fungi live together in a mutualistic relationship. The constituents of lichen — algae, bacteria and fungi — can arrive in the form of spores with wind or water. These spores, when they land on the surface, will start the colonisation of these rocks. It is important to note here that we may also have seeds of trees coming together with the spores. However, in the absence of soil, the trees (and their saplings) cannot be supported, and will die off due to exposure to the elements. Only the spores of lichens will be able to colonise the rock, for they are hardy and require simple resources.

The first form will look much like rust on iron. This is known as the crustose lichen, and often appears yellow-red in colour [Fig. 3.5b]. When crustose lichens grow, the filaments of the fungi are able to enter the fine fissures on the rock surface. When these fungi die, the decomposition of their bodies produces acids and gases which further weaken the rock. The presence of life forms and their activity thus increases the rate of disintegration of the rock.

After a while, we'll start observing foliose lichen [Fig. 3.5c]. These are leaf-like in structure, and can out-compete the crustose lichen. On several rocks, we may observe both crustose and foliose lichen in different areas [Fig. 3.5d].

The action of lichens further breaks the rocks, and makes it suitable for the growth of moss [Fig. 3.5e]. Mosses are non-vascular flowerless plants. They further break the rock down, and we now observe a thin layer of soil forming. This soil is comprised of minerals from the disintegrating rock, together with the organic matter from the lichens and the moss.

The presence of soil now permits grasses to grow [Fig. 3.5f]. Grasses are flowering plants. They further change the soil, and after a while, we start observing shrubs, which are perennial woody plants [Fig. 3.5g,h] with multiple stems and medium height. They then get replaced by forests with trees [Fig. 3.5i], which is the climax community, meaning that it is the best-adapted community and will not be replaced by other seres.

A common theme running through different seral communities is that each community makes changes to its environment (such as by converting rocks into soil), and in the process constructs conditions where another community is able to out-compete the original community. It is not necessary that all of the community gets replaced *in toto* — we may also observe forests with shrubs, grasses and even lichens [Fig. 3.5j] — often at one place. But the dominant community does change during ecological succession.

We thus observe a continuous transition from a pioneer community to the climax community. The hardy species which establish themselves in a disrupted (or new) ecosystem and trigger the process of ecological succession are known as pioneer species. They make the pioneer community, and are characterised by

1. ability to grow on bare rocks, nutrient-poor soil or water,

2. ability to tolerate extreme conditions such as heat and cold,

3. having less nutritional requirements — they are typically photoautotrophs,

4. small size,

(a) Bare rocks.

(b) Crustose lichen in Dachigam National Park.

(c) Foliose lichen in Mukurthi National Park.

(d) Crustose and foliose lichen on a rock in Mukurthi National Park.

(e) Moss on a boundary pillar.

(f) Herbaceous stage: grasses in Kedarnath Wildlife Sanctuary.

(g) Shrub stage seen in Wild Ass Sanctuary.

(h) A close-up of a shrub.

(i) Forest / woodland stage: forests in Manali.

(j) Crustose lichen, foliose lichen and moss on a tree bark in Mukurthi National Park. Herbs, shrubs and the tree are also visible.

Figure 3.5: An example of stages in ecological succession.

5. short life span with rapid growth — many pioneer species are annual species,

6. ability to disperse through spores or seeds, and

7. prolific seed production.

On the other extreme is the climax community — a biological community of plants, animals and fungi which, through the process of ecological succession in the development of vegetation in an area over time, have reached a steady state. Climax communities are characterised by

1. having vegetation tolerant of environmental conditions,

2. high species diversity,

3. having well-formed spatial structure,

4. complex food chains that provide stability,

5. having an equilibrium between gross production and respiration, and between uptake and release of nutrients,

6. having a species composition that continues for a long time, and

7. being good indicators of the climate and other conditions of the area.

Four main kinds of climax are recognised:

1. Climatic climax: controlled by the climate of the region

2. Edaphic climax: controlled by the soil conditions of the region

3. Catastrophic climax: controlled by some catastrophic event such as wildfire

4. Disclimax: controlled by some disturbance (man or domestic animals)

Three kinds of succession can be observed

1. Primary succession: "Successional dynamics beginning with colonisation of an area that has not been previously occupied by an ecological community." These are often observed on newly exposed rock or sand surfaces, rocks arising out of lava flows and on newly exposed glacial tills.

2. Secondary succession: "Successional dynamics following severe disturbance or removal of a pre-existing community." These are often observed, say, after events such as forest fires.

3. Cyclic succession: "Periodic changes arising from fluctuating species interactions or recurring events."

Secondary and cyclic succession are often faster than primary succession due to several reasons:

1. Soil is already formed, reducing the time required.

2. Spores and seeds are already present in the soil as seed and spore banks, which makes colonisation faster.

3. Regeneration of some plants from roots can occur.

4. Soil fertility is typically high enough to support organisms in a short period of time.

Some examples of succession are:

1. Lithosere primary succession: Rock → Crustose lichen stage → Foliose lichen stage → Moss stage → Herbaceous stage → Shrub stage → Woodland stage → Climax stage

2. Hydrosere primary succession: Water → Phytoplankton stage → Submerged stage → Floating stage → Reed swamp stage → Sedge-meadow stage → Woodland stage → Climax stage

3. Secondary succession: Forest → Forest fire → Forest incompletely destroyed → Herbaceous stage → Shrub stage → Woodland stage → Climax stage

Clements [Clements, 1916] has recognised the following phases of succession:

1. Nudation: Succession begins with the development of a bare site, called nudation (disturbance).

2. Migration: It refers to the arrival of propagules such as seeds or spores.

3. Ecesis: It involves the establishment and initial growth of the vegetation.

4. Aggregation: Increase in numbers and population densities of organisms.

5. Competition: As vegetation becomes well established, grows, and spreads, various species begin to compete for space, light and nutrients.

6. Reaction: During this phase autogenic changes (changes brought by the organisms themselves) such as the buildup of humus affect the habitat, and one plant community replaces another.

7. Stabilisation: A supposedly stable climax community forms.

Three main theories of climax are recognised:

1. Monoclimax or Climatic Climax Theory — advanced by Clements in 1916: It states that there is only one climax whose characteristics are determined solely by climate. The processes of succession and modification of environment overcome the effects of other factors such as topography, parent material of the soil, etc.

2. Polyclimax Theory — advanced by Tansley in 1935: It states that the climax vegetation of a region consists of more than one vegetation climaxes controlled by soil moisture, soil nutrients, topography, slope exposure, fire, and animal activity.

3. Climax Pattern Theory — advanced by Whittaker in 1953: It states that there is a variety of climaxes governed by responses of species populations to biotic and abiotic conditions. The nature of climax vegetation will change as the environment changes, with the central and most widespread community being the climatic climax.

Different communities may be explained by one of these theories.

3.3 CHANGES IN POPULATIONS IN A COMMUNITY

3.3.1 Exponential growth

Let us understand population growth with a thought experiment. Consider an island on which 10 individuals of a species — say rats — are translocated. We'll call it generation 0. There are no predators, and there is no dearth of food or other resources. Thus, the population grows.

We define R_0 as the net reproductive rate. It is the number of female offsprings produced per female per generation, and gives the rate at which the population grows. Let us consider a small R_0 of 1.5. It would mean that every generation will be 1.5 times the previous generation. Thus, the first generation will have $10 \times 1.5 = 15$ individuals, the second generation will have $15 \times 1.5 = 22.5$, or 23 individuals, and so on. Mathematically, we can represent the growth in population as

$$N_{t+1} = R_0 \times N_t$$

where

N_t = Population size at generation t
N_{t+1} = Population size at generation $t + 1$
R_0 = Net reproductive rate = Number of female offsprings produced per female per generation

The population growth over several generations is depicted in table 3.1 and in figure 3.6a. We've made the simplification that there are no deaths in this population. In actuality, old individuals of the previous generations will die, but considering the large, and increasing size of each successive generation, we choose to neglect those deaths, as an approximation.

Table 3.1: Increase in population for $N_0 = 10$ and constant $R_0 = 1.5$.

Generation	Population size
0	10
1	15
2	22.5, or 23
3	33.75, or 34
4	50.625, or 51
5	75.9375, or 76
6	113.90625, or 114
7	170.859375, or 171
8	256.2890625, or 256
9	384.43359375, or 384

We can observe that the population grows very fast, and in a short time the population is several times the original population. This is exponential growth.

3.3.2 Logistic growth

In practice we rarely observe such exponential growths for long periods. This is because our assumptions — that there is no dearth of food or other resources — do not stand for long periods. The space on the island is finite. The primary production — and thus the food supply — is finite. In a short period, the population will increase to such a large extent that several resources — food, water, shelter, space, etc. — will start getting overwhelmed. There will soon be a shortage of resources. And individuals will have to compete against each other to get the resources.

We define carrying capacity as the maximum number of individuals that an area can support

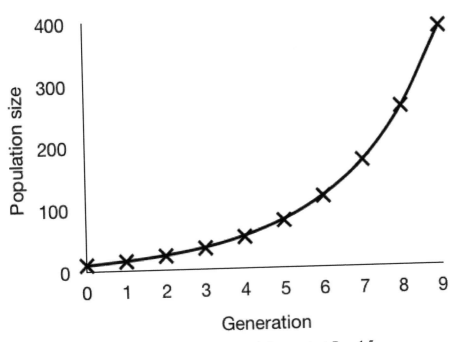

(a) Exponential population growth for constant $R_0 = 1.5$.

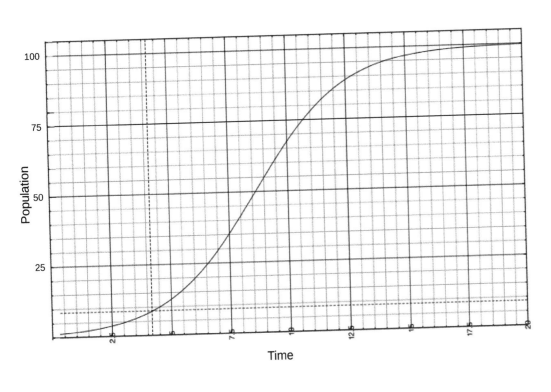

(b) Logistic population growth for $K = 100$ and $r = 0.533$.

Figure 3.6: Graphs depicting population growth.

without degradation. As the population starts reaching the carrying capacity, the shortage of resources ensures that the rate of population growth decreases — because of less individuals being born, or because of larger number of deaths. At the carrying capacity the rate of population growth effectively becomes zero, or else the habitat will start to degrade, reducing the carrying capacity even further. Thus, R_0 is not a constant in real life. It varies with the population size.

Mathematically, we can represent this by the logistic growth equation:

$$\frac{dN}{dt} = rN \times \left(\frac{K-N}{K}\right)$$

where

N = Population size at time t

K = Carrying capacity of the environment

r = intrinsic growth rate

When $K \gg N$, $\frac{K-N}{K} \to 1$, and we have

$$\frac{dN}{dt} = rN$$

Thus, the rate of growth is proportional to the population present at any time. When $N \approx 0$, $\frac{dN}{dt} \approx 0$, and the population grows at a very slow rate. We call this stage — where N is very small — the lag phase.

As N increases, $\frac{dN}{dt}$ increases — and there is an exponential growth. We call this stage — where N is substantial and the population grows exponentially — the log phase.

When N increases so much that it becomes close to K ($N \approx K$), $\frac{K-N}{K} \to 0$, and we have

$$\frac{dN}{dt} = 0$$

Thus there is no further growth in population. The population becomes constant. We call this stage — where N is close to K and the population growth is close to zero — the stationary phase.

These phases are represented in figure 3.6b.

3.3.3 Lotka-Volterra equations

Another factor regulating the growth of population is the presence of predators and the availability of prey. We may represent predator-prey relations through Lotka-Volterra equations:

$$\frac{dV}{dt} = rV - \alpha VP$$

$$\frac{dP}{dt} = \beta VP - qP$$

where

V = prey population at time t

P = predator population at time t

α, β, r, q = constants

Essentially, what the Lotka-Volterra equations say is that the rate of increase in prey population will depend upon the number of individuals in the prey population (since more individuals $\implies$ more births), but reduced by a factor given by the product of a constant with prey and predator populations. In the absence of the predator population ($P = 0$), the prey population will essentially show an exponential growth and degrade the habitat. (This also explains why we need predators such as tiger to conserve our forests.)

The rate of increase in predator population will not only depend upon the number of individuals in the predator population, but also those in the prey population (which provide food), and reduced by the product of a constant with the predator population (since larger predator population size $\implies$ more deaths).

If tiger and chital populations are governed by Lotka-Volterra dynamics (with $r = 0.008$, $q = 0.8$, $\alpha = \beta = 0.001$), and the initial population sizes are given by:

Tigers, $P = 14$, and

Chital, $V = 1,000$, we get the changes in tiger and chital population as depicted in figure 3.7.

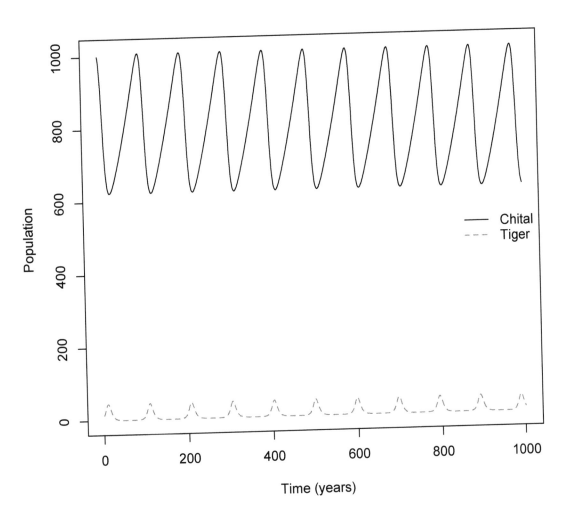

Figure 3.7: Plot of chital and tiger populations with the given initial conditions.

We can observe from the graph that the chital and tiger populations show cyclical oscillations. This is expected since

1. Large prey population and small predator population $\implies$ Predators get ample food and increase their population, while prey population reduces as the predator population increases.

2. Small prey population and large predator population $\implies$ Predators have shortage of food, increasing competition and reducing their population growth rate, while prey population increases as the predator population decreases.

Thus, in the short-term, the model predicts that the chital population shall decline, while the tiger population shall increase. This would then be continued with a cyclical steady state, where the populations of chital and tiger go through ups and downs, with the tiger population at a lag with respect to the chital population. This is because the tiger population will utilise chital population as prey to increase their numbers while decreasing the chital numbers. After some time, the chital population will not be sufficient to support the large tiger population, and the tiger population will start decreasing. With less number of predators, the chital population will start increasing and re-establish itself. Thus we reach the same starting conditions of large chital population and low tiger population. And this cycle will continue.

Thus the predator-prey relationship ensures that the two populations regulate each other. Lotka-Volterra equations can also be applied to other relationship situations, such as where two populations are competing against each other. But that is beyond the scope of this book.

CHAPTER 4

Threats to wildlife resources

Haematopinus oliveri, also known as the pygmy hog-sucking louse, is a critically endangered species of insect. It is a parasite that lives off the pygmy hog (*Porcula salvania*, Fig. 4.1), a species of hog that itself is endangered. The pygmy hog is a small species of mammal, measuring only up to 70 cm in length and 30 cm in height. In the past, it was found over large areas. Being a species of the grasslands, it could be found throughout the foothills of the Himalayas, from Uttarakhand in India, through Nepal and Bhutan, and up to Assam. These areas were characterised by a continuous stretch of wet alluvial grasslands, called *terai*. There was little pressure from humans, since these wet areas were also badly infested with mosquitoes and malaria. With the advent of modern insecticides that controlled the mosquito population, huge swaths of these alluvial grasslands were converted into agricultural fields bearing sugarcane, paddy and wheat. This not only fragmented the habitat of the pygmy hog, but also reduced its area, with the result that out of around 2,450 square kilometre of estimated extent of occurrence, today the pygmy hog has an estimated area of occupancy of only 20 square kilometre ($< 1\%$). The current estimates state that only 100–250 animals exist in the wild today. Such rarity often increases the demand for species as an exotic pet or a specimen for private collections, incentivising poachers to extract the few remaining individuals, pushing it towards complete extinction. And with the pygmy hog population declining, the populations of the species dependent upon the pygmy hog (say for food, pollination, dispersal, etc.) are also decreasing. *Haematopinus oliveri* in particular has decreased so much that it has now become critically endangered. This example illustrates several facts. First, that species are localised. They are only found in certain locations (such as the pygmy hog being found only in the grasslands in the Himalayan foothills) and not all locations. Second, if these specific locations are harmed or diverted in some ways, the existence of the species that are localised in these areas comes under threat. Third, the rarer a species is, the greater is its demand by humans for collections, which vastly increases the threat of poaching or removal. Fourth, if a species is under threat, those species that are dependent upon the threatened species will also come under threat. This understanding has important ramifications in explaining the threats to our wildlife resources. Threats arise because species are localised, and the localisation areas are being threatened. This threat emanating from the destruction of habitat adds to the threat emanating from the removal of individuals of the species through hunting and capture, and harms not just one species, but an interdependent set of species. We now explore these points in greater detail.

4.1 DISTRIBUTION AND ABUNDANCE OF WILDLIFE RESOURCES

Different organisms are found in different locations on Earth. Polar areas have polar bears, camels are found in deserts and elephants roam the forests. Biogeography is the scientific discipline that studies the geographical distribution of life on Earth and the reasons for the patterns one observes on different continents, islands and oceans. It asks the following questions: where is a species found (actual range)? What are the suitable areas where the species *may be* found (potential range)? And

Figure 4.1: Pygmy hog is an animal found in Manas Tiger Reserve.

why is it that the species is found in the areas where it is found (causal factors for distribution)? The discipline originated from observations of several naturalists such as Charles Darwin who noted that different areas on the Earth have vastly different organisms.

An extract from "The Voyage of the Beagle" by Charles Darwin [Darwin, 2012]

"Two kinds of geese frequent the Falklands. The upland species (*Anas Magellanica*) is common, in pairs and in small flocks, throughout the island. They do not migrate, but build on the small outlying islets. This is supposed to be from fear of the foxes: and it is perhaps from the same cause that these birds, though very tame by day, are shy and wild in the dusk of the evening. They live entirely on vegetable matter. The rock-goose, so called from living exclusively on the sea-beach (*Anas antarctica*), is common both here and on the west coast of America, as far north as Chile. In the deep and retired channels of Tierra del Fuego, the snow-white gander, invariably accompanied by his darker consort, and standing close by each other on some distant rocky point, is a common feature in the landscape."

The range or distribution of a species is defined as the geographical area within which that species can be found. These geographical areas where a particular species is found are characterised by certain specific properties that make it suitable for the species — a habitat for the species. Habitat is defined as the "subset of physical and biotic environmental factors that permit an animal (or plant) to survive and reproduce [Block and Brennan, 1993]." It is a species-specific concept associated with a geographical location.

Habitats are characterised by physical factors such as soil, moisture, temperature and light intensity, and biotic factors such as vegetation, availability of food and presence or absence of predators. Temperature and precipitation are two important factors that govern most of the properties of the habitats.

Different places on the Earth have different habitats, and different residents living in these habitats. Alpine meadow habitats [Fig. 4.2a] have vast spans of grasslands in low temperature and windy areas. Alpine forest habitats [Fig. 4.2b] have low temperature forests, typically teeming with coniferous species. Deciduous forests are characterised by trees that shed their leaves in particular seasons, typically the dry season. This is an adaptation to conserve moisture, and deciduous forests are found in areas with scarcity of moisture in certain seasons of the year. Cooler, moist areas have moist deciduous forests [Fig. 4.2c] such as sal forests, and drier areas have dry deciduous forests [Fig. 4.2d] such as teak forests. When moisture becomes very less, we observe a transition to scrub forests [Fig. 4.2e,f] — open forests with shrubs, thorny vegetation and grasses — and finally to desert habitats [Fig. 4.2g,h] covered with sand. At the other extreme are the equatorial forests found in places with hot and moist climate [Fig. 4.2i]. Aquatic habitats include fresh water habitats such as ponds [Fig. 4.2j], lakes [Fig. 4.2k] and rivers [Fig. 4.2l], brackish water habitats such as estuaries [Fig. 4.2m] and lagoons [Fig. 4.2n], and marine habitats [Fig. 4.2o]. Some areas witness cyclic shifts between terrestrial and aquatic habitats. Good examples are floodplains [Fig. 4.2p], sea coasts [Fig. 4.2q], intertidal zones [Fig. 4.2r], and agricultural fields under inundated agriculture (such as paddy fields, Fig. 4.2s). Specialised habitats include complexes of grassland and forests such as Shola forests [Fig. 4.2t], bogs [Fig. 4.2u], fynbos, mangroves [Fig. 4.2v], salt extraction sites [Fig. 4.2w] and wastewater treatment areas [Fig. 4.2x] that have unique biodiversity.

4.1.1 Suitability of habitats for different organisms

Which species will be found in a particular habitat will depend largely on whether the habitat meets the needs of the species. Organisms have a need for space, food, water and safety. If a habitat provides these, the organisms may be found there. It will also depend on the adaptations that the species has evolved over time to make the best use of the habitat they are found in. Thus camels — which live in hot, dry, resource-scarce and sandy environs of deserts — have adaptations such as padded feet for walking on the hot sand, large eyelashes to protect their eyes from blown sand, a hump to store fat and water, and thick skin to protect them from heat. While many species will find it extremely difficult to survive in the hot and dry desert conditions, camels can happily make use of the harsh habitat because they are uniquely adapted for these habitats. But these very adaptations that help them in deserts may become an incumbrance in more moderate conditions. When there is little need to carry a supply of food and water, the added weight will slow the camels down in comparison to other species of the moderate habitats. And then they'll become easy prey to the predators found in the moderate habitats, and, in essence, will get 'out-competed' by those animals that have adaptions suited for the moderate conditions. This explains why we don't find camels in thick forests with moderate conditions, even though these forests have plenty of food, water and space for the camels. Similarly, snow leopards will be found in high mountainous regions with cold climates and rugged terrains because they are best adapted to living in those cold areas with rugged terrains, and perhaps better adapted than any other predator of their niche. It is important to highlight here that there are no 'right' or 'wrong' adaptations. The very adaptations that make it possible to efficiently utilise one habitat (the 'right' adaptations for that habitat) may increase inefficiency in another habitat (thus becoming 'wrong' adaptations for that habitat). This is why the variety of habitats found on our planet has translated into the variety of life forms — the biodiversity of our planet.

What makes a particular habitat suited for an organism can be learnt through observations and experiments. We can begin by noting the locations where the species is found, and the locations where the species is not found. Next we can make a list of the attributes of those locations, and try to correlate the attributes with the presence or absence of the species. This may be helped by observing and investigating the adaptations found in the species, and confirmed through experiments. An early example can be seen in "The structure and distribution of coral reefs" by Charles Darwin where he describes the relationship between the presence of movable sediments and the growth of corals [Darwin and Bonney, 1889]:

"the recoil of the sea from a steep shore is injurious to the growth of coral, although waves breaking over a bank are not so. Ehrenberg also states, that where there is much sediment, placed so as to be liable to be moved by the waves, there is little or no coral; and a collection of living

(a) Alpine meadows in Dachigam National Park.

(b) Alpine forests in Uttarakhand.

(c) Moist deciduous forest with sal trees in Timli Range.

(d) Dry deciduous forest with teak trees in Harda division of Madhya Pradesh.

(e) Scrub forest in Ranthambhore Tiger Reserve.

(f) Scrub forest with shrubs in Wild Ass Sanctuary.

(g) Sand dunes in Rajasthan

(h) support animals like the spiny tailed lizard, seen here in the Desert National Park.

(i) Equatorial forests, seen here in the Andamans, are characterised by high temperatures and heavy rainfall.

(j) A pond in Nauradehi Wildlife Sanctuary.

(k) The Dal lake in Srinagar.

(l) The Dhauli Ganga river in Uttarakhand is a specialised habitat.

(m) Estuaries are characterised by mixing of fresh and saline water.

(n) Chilika lake is an example of a wetland habitat.

(o) The marine habitat has saline water.

(p) The floodplains of Brahmaputra have lush grasses, as seen here in Kaziranga Tiger Reserve.

(q) The sea coast in Andaman islands.

(r) The intertidal zone in Marine National Park, Jamnagar.

(s) Inundated agricultural fields are an example of man-made wetlands.

(t) The Shola forests of Coorg are a unique habitat.

(u) A bog in Finland.

(v) Mangrove forests are a unique habitat supporting large biodiversity.

(w) Salt extraction sites such as this in the Little Rann of Kutch also serve as wetlands.

(x) Wastewater treatment areas, such as the Kakreta facility serve many functions of wetlands.

Figure 4.2: Some examples of the variety of habitats.

specimens placed by him on a sandy shore died in the course of a few days. An experiment, however, will presently be related, in which some large masses of living coral increased rapidly in size, after having been secured by stakes on a sand-bank. That loose sediment should be injurious to the living polypifers, appears, at first sight, probable."

We can use a similar technique for other habitats. When we observe that certain hill slopes have vegetation, and others are devoid of vegetation [Fig. 4.3], we can analyse the differences in these slopes. Perhaps vegetation is found on gentle slopes but not on steep slopes; or vegetation may be found on moist slopes but not on drier slopes; or there may be a difference in the amount of sunlight being received (Southern slopes in Northern hemisphere and Northern slopes in Southern hemisphere get more Sun); or the vegetation are responding to the speed of wind. We may observe the adaptations of the plants. Are these plants adapted for low moisture (waxy leaves, spines, thorns, recessed stomata, large roots, etc.), or for high wind speeds (short height, wavy leaves, strong stem that bends, etc.), or for snow (coniferous leaves that permit snow to fall down)? That might give a clue. And to confirm, we may perform an experiment. If perhaps low moisture is the reason for a denuded slope, artificial irrigation should result in vegetation. If high wind speeds are preventing the growth of plants, then perhaps a windbreak may permit plants to come up. And so on.

Figure 4.3: This picture of the Shivalik Hills depicts certain hill slopes with vegetation, and certain hill slopes devoid of vegetation.

Through such experiments, it has been found that several physical and chemical factors simultaneously govern treelines on mountains [Parker, 1963]. These factors include:

1. lack of soil,

2. drought,

3. desiccation of leaves in cold winter,

4. lack of snow, exposing plants to winter drying,

5. excessive snow, lasting through the summer,

6. short growing season,

7. rapid heat loss at night,

8. excessive soil temperatures in the day, and

9. mechanical aspect of high winds.

In this way, we can list out the *push and pull factors* of the habitat for a species. Push factors are conditions that drive organisms away from an area. Examples of push factors include food scarcity, inhospitable climate and presence of predators, which 'push' the organisms away from the area. Pull factors are conditions that attract organisms to an area. Examples of pull factors include availability of food and water, amiable climate and absence of predators. In these situations, the organisms will thrive well in the habitat, and we will say that the species gets 'pulled' to the habitat.

The impacts of different push and pull factors are determined by two laws:

1. Liebig's law of the minimum: It states that "the rate of any biological process is limited by that factor in least amount relative to requirement, so there is a single limiting factor." Thus, if a habitat provides 10% requirement of nutrient A, 20% requirement of nutrient B, 40% requirement of nutrient C and 80% requirement of nutrient D, the growth of the species will be regulated by the availability of nutrient A, since nutrient A is available in the least amount relative to the requirement. And if we add nutrient A experimentally, we will find a large response to the growth of the species.

2. Shelford's law of tolerance: The geographical distribution of a species will be controlled by that environmental factor for which the organism has the narrowest range of tolerance. Thus, if a species can tolerate 10% to 100% humidity levels, but temperatures only from 32 °C to 34 °C, then its geographical distribution will be controlled not by humidity — to which the organism has wide tolerance, but by temperature — for which it has a narrower range of tolerance.

This has important ramifications in the context of global warming. The distribution of several insects is regulated by the ambient temperature. This is especially evident in mountainous regions where insects are not found in greater heights where temperatures are low. But with increasing temperatures, we are witnessing gradual increase in the median heights where different insect species are seen [Siraj et al., 2014, Awadhiya, 2018] [Fig. 4.4a]. Since many insects also serve as vectors for diseases, their increased presence in these locations will expose people and animals to diseases that they were hitherto largely protected from. At the same time, what will happen to the species that live on the top of these mountains? These species, with unique adaptations that make them survive in the harsh cold climates, will be unable to compete in the changed, warmer conditions, and will die off. In essence, with global warming, we are witnessing an 'escalator to extinction,' where the species are moving to higher and higher altitudes, and towards their extinction [Fig. 4.4b].

Together with the physical factors such as temperature and moisture, several biological factors also govern the distribution of organisms. They include

1. Allelopathy: Allelopathy is the chemical inhibition of one organism by another through release of germination inhibitors or growth inhibitors. Good examples include secretion of antibiotics that kill microorganisms.

(a) With rising temperatures, changes in the range of organisms such as insects is expected.

(b) The escalator to extinction.

Figure 4.4: Some examples of changes due to global warming.

Allelopathy is easily observed in teak forests [Fig. 4.2d] where the forest floor is largely devoid of vegetation due to the release of allelopathic chemicals from the fallen leaves of teak trees. It can be experimentally demonstrated by adding an extract of teak leaves to germinating and young plants of other species. Inhibition of germination and growth is observed [Macias et al., 2000, Kole et al., 2011].

2. Predation: Predation is the preying of one organism on others. When predation is the causal factor in regulating distribution of prey organisms, then experimental removal of the predator will lead to a rapid increase in the prey population, and the prey will establish in those areas where previously it was not found due to the activity of the predator. An example is the experimental removal of sea urchin (*Paracentrotus*, predator) which brings cover of algae (prey) in hitherto blank areas [Fig. 4.5a].

3. Presence of prey (prey governing the distribution and abundance of the predator): In certain cases, the predator is so dependent upon the prey that it is only found in areas where the

prey is found. Good examples include the pygmy hog-sucking louse (*Haematopinus oliveri*) which is only found where the pygmy hog is found, and the fly *Drosophila pachea*, whose breeding is dependent upon the chemical sterol $\Delta^7 - Stigmasten - 3\beta - ol$ that is found in its prey, the senita cactus (*Lophocereus schottii*). Thus *Drosophila pachea* is only found where senita cactus (*Lophocereus schottii*) is found [Heed and Kircher, 1965].

4. Interspecific competition: In certain cases, while two species are not in a predator-prey relationship, they use the same resources, and may aggressively defend those resources. In such cases, one species may out-compete the other species and displace it from certain habitats. An example is the displacement of the red-winged blackbird (*Agelaius phoeniceus*) by the tricoloured blackbird (*Agelaius tricolor*) [Orians and Collier, 1963]. Both these species use the same resources, and the tricoloured blackbirds — through use of aggression — outcompete and displace out the red-winged blackbirds from their colonies [Fig. 4.5b].

Along with physical and biotic factors, other factors may also be responsible for the observed abundance and distribution of organisms. Some such factors are

1. Behavioural factors: Some organisms have innate and learnt 'preferences' when it comes to selecting a habitat. Thus, even though multiple locations may fulfil the requirements of the species, individuals of the species will prefer one location and not the other. This 'habitat selection' may be defined as "a hierarchical process of behavioural responses that may result in the disproportionate use of habitats to influence survival and fitness of individuals [Jones, 2001]." For example, chipping sparrows spend a majority of their time (around 71%) in pine trees, and only around 29% of their time on oak trees [Klopfer, 1963].

2. Dispersal: Dispersal is defined as "the movement of individuals away from their place of birth or hatching or seed production into a new habitat or area to survive and reproduce." It is possible that even though a habitat has adequate pull factors, and is also a preferred habitat for a species, the species is not found there just because it has not reached the place to begin the process of colonisation.

3. Migration: Migration is "the regular, seasonal movement of animals, often along fixed routes [Fig. 4.6]." This is either to obtain better resources (e.g. food, breeding sites) or to shift from harsh to a more amiable climate. A species may not be found in a suitable preferred habitat because it has migrated out, and is yet to return.

4. Anthropogenic, or man-made factors such as clearing of forests, poaching and pollution: Even though a habitat might be suitable and preferred for a species, the species might have been hunted out or displaced due to human activity.

4.1.2 Transplant experiments

To pinpoint the cause of observed distributions, we may perform transplant experiments [Fig. 4.7a]. In these experiments, organisms are artificially moved ('transplanted') from an area where they are found to another area where they are not found. If the organism is able to establish and thrive in the new location ('transplant successful'), then it may be interpreted that the distribution is limited either because:

1. the area is inaccessible (physical barrier), or

2. time has been too short to reach the area (dispersal time insufficient), or

3. the species fails to recognise the area as a suitable living space (habitat preference).

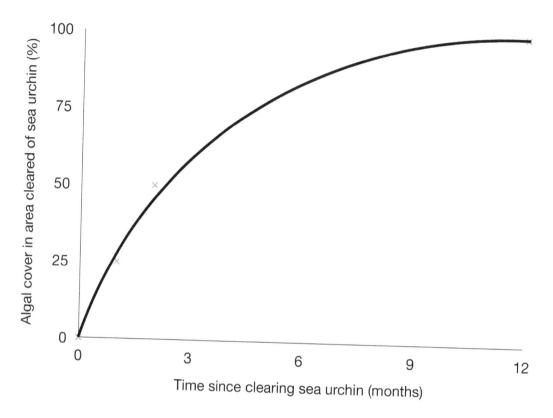

(a) Increase in prey population after removal of predator. Removal of sea urchin leads to recolonisation by algae. Data source: [Kitching and Ebling, 1961].

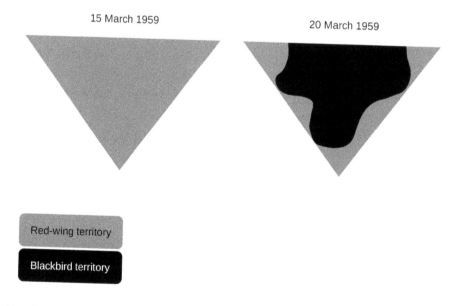

(b) Displacement of red-winged blackbirds by tricoloured blackbirds.

Figure 4.5: Some examples of biological factors regulating distribution and abundance.

(a) Demoiselle cranes are migratory birds.

(b) Demoiselle cranes from Mongolia and China spend the winter in the Indian subcontinent. This is a flock in Jodhpur, Rajasthan.

Figure 4.6: An example of migratory species: demoiselle crane.

On the other hand, if the organism is unable to establish and thrive in the new location ('transplant unsuccessful'), then it may be interpreted that the distribution is limited because of:

1. other species (predation, parasitism, competition, etc.), or

2. physical and chemical factors.

Modifications to the new site (e.g. provisioning of protective cages, wind breaks, irrigation, etc.) may be used to further pinpoint the governing factors.

The control experiment [Fig. 4.7a] consists of transplanting organisms within their areas of distribution to ensure that unsuccessful transplants are not artefacts of faulty experimentation (such as damage to plants or death to animals during the process of transplanting).

In this way, we can ascertain the causes of distribution and abundance of organisms [Fig. 4.7b].

4.2 KINDS OF THREATS TO SPECIES

From the preceding section, we can say that a species is found in those areas that it finds suitable — in other words, those areas that 'pull' the species towards them. And a species avoids those areas from which it is 'pushed' out. If it so happens that a species is 'pushed' from everywhere and does not find any areas that 'pull' it, it would have no place to go to. And that would be a threatening situation for the survival of the species, and the species might be pushed towards extinction.

4.2.1 Declining population paradigm and small population paradigm

The factors that push a species towards extinction can be divided into two categories [Caughley, 1994]:

1. factors pushing a population towards smaller numbers through population dynamics: called the *Declining population paradigm*. These include

    (a) No suitable habitat:
        i. too hot, too cold
        ii. no trees, no food, no nutrients
        iii. completely burnt out
        iv. rich in noxious factors: too polluted
        v. not suited behaviourally: habitat selection at play
    (b) Competed out:
        i. invasive species
        ii. too many predators or diseases
    (c) Killed out:
        i. heavy poaching

2. factors pushing a small population towards extinction: called the *Small population paradigm*. These include *small-population dynamics*:

    (a) Allee effect: the positive correlation between population density and individual fitness. Examples include cooperative hunting. Animals such as wolves and wild dogs hunt cooperatively. If the pack size goes below a threshold, the pack will be unable to hunt efficiently, and the fitness of all the individuals will go down. Another example is cooperative defence as in bee hives, where a low population will be unable to mount a sufficient defence, and the hive may get raided. Yet another example is search for

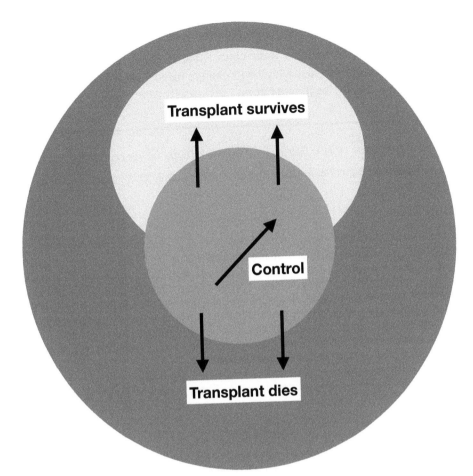

(a) Transplant experiments are used to investigate the observed distribution of organisms and to identify their potential range.

DICHOTOMOUS KEY

(1) Is species absent because area is inaccessible: YES → DISPERSAL
(1) Is species absent because area is inaccessible: NO
If NO then
(2) Is the habitat an unpreferred habitat for the species: YES → HABITAT SELECTION
(2) Is the habitat an unpreferred habitat for the species: NO
If NO then
(3) Are other species causing predation, parasitism, competition or diseases: YES → INTER-SPECIFIC INTERACTIONS
(3) Are other species causing predation, parasitism, competition or diseases: NO
If NO then
(4) Species absent because of physical or chemical factors

(b) A dichotomous key to understand the absence of species from a location.

Figure 4.7: Transplant experiments and a dichotomous key.

mates. If the population density is less, each individual will have great difficulty in searching for mates, and may have to traverse large areas to perform the search, with low rates of success.

(b) Chance deaths due to environmental variations, catastrophes, diseases, etc.

(c) Stochastic sex ratio: there is a finite chance that all offsprings are born male, or that all offsprings are born female, which will change the sex ratio and have repercussions on the birth rates for the next generation.

How real is the threat? Glimpses from Biogeography

According to the island biogeography model [Wilson and MacArthur, 1967], species richness, S of an island is given by

$$S = C \times A^z$$

where A is the size of the island

C, z are constants depending on the set of species and the island

In other words, the theory suggests that the species richness depends upon the area of the island: the more the area, the more the species richness. This is expected because larger areas typically have more number of habitats, which leads to a higher species richness. At the same time, larger areas can support large populations of species, which protects them from extinction. Larger areas are also less vulnerable to catastrophes, since smaller catastrophes will not impact the whole area.

However, the species richness is not directly proportional to the area of the island, but depends upon some power of the area, z, which needs to be determined.

Through measurements, z has been found to lie between 0.15 and 0.35.

To estimate the rate of species loss, let us take $z = 0.30$. Then, for an area A_1, the species richness is given as

$$S_1 = C \times A_1^{0.30}$$

What will happen when the area decreases due to loss of habitat? Let us calculate the scenario where the area decreases by 90%, i.e. only 10% of the original habitat remains:
$A_2 = 0.1 \times A_1$
Then,

$$S_2 = C \times (0.1 \times A_1)^{0.30}$$

This gives

$$\frac{S_2}{S_1} = \frac{C \times (0.1 \times A_1)^{0.30}}{C \times A_1^{0.30}}$$

$$\implies \frac{S_2}{S_1} = 0.1^{0.3}$$

$$\implies \frac{S_2}{S_1} = 0.5012 \approx 50\%$$

Thus, $S_2 = \frac{1}{2} \times S_1$

So, by reducing area by 90%, the species richness becomes halved. This is expected because while those organisms that require a large home range will get extinct, many organisms with smaller home range requirements will probably survive.

How does this correlate with the real world?

The rate at which tropical forests are actually decreasing is $\approx 1.8\%$ per annum. Let us see what this means. On taking the most conservative value of z (0.15), this would translate to an annual loss of 0.27%.

The total number of species is unknown, and most species remain undocumented. Scientists have estimated around 30 million species of tropical arthropods alone [Erwin, 1982]. However, for the present, let us take a very conservative estimate of around 10 million species in tropical forests.

Then the annual loss of species from tropical forests is given by

10,000,000 × 0.27/100
= 27,000 species per year
And this is the most conservative estimate! Most of the species are getting lost even without being documented! The situation really is too grim.
Similarly, we may estimate the losses in other ecosystems.

The declining population paradigm factors and the small population paradigm factors emanate from population dynamics. Two kinds of factors operate on a population at all times:

1. deterministic factors — acting at large population sizes, and including variables such as

   (a) birth rate,

   (b) death rate, and

   (c) population structure.

2. stochastic factors — important when the population sizes are smaller, and including variables such as

   (a) demographic stochasticity including occurrence of probabilistic events such as reproduction, litter size, sex determination, and death,

   (b) environmental variation and fluctuations,

   (c) catastrophes such as forest fires and diseases,

   (d) genetic processes including loss of heterogeneity and inbreeding depression,

   (e) density dependent mortality on exceeding the carrying capacity of the habitat, and

   (f) migration among populations.

The important factors driving a species towards extinction may be remembered with the mnemonic HIPPO:

1. H: Habitat loss

2. I: Invasive species

3. P: Pollution

4. P: human over-Population

5. O: Over-harvesting

When these factors operate, the population may be subjected to the extinction vortex [Fig. 4.8], which can be defined as "the process that declining populations undergo when a *mutual reinforcement occurs* among biotic and abiotic processes that drives population size downward to extinction."

In other words, the declining population creates conditions that further decline the population in a positive feedback loop through processes such as inbreeding depression, demographic stochasticity and genetic drift. Together, these make a small population even smaller. And with a smaller population, the processes including inbreeding depression, demographic stochasticity and genetic drift become even stronger. This process goes on till the population becomes extinct.

This also suggests that those species that have a small population are at a much higher risk of extinction. Often, rarer species have small populations, because of which they are rare. Hence the susceptibility to extinction is a function of the rarity of the species — the rarer the species, the higher its chances of getting extinct. But then, why are some species rare? It turns out that rarity is a function of the ecology and evolutionary characteristics of the species. There are three natural reasons why some species are rare:

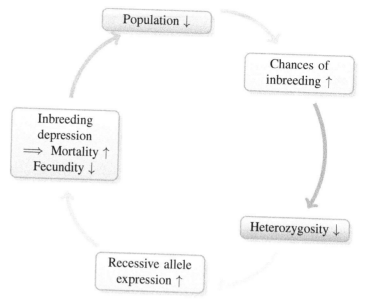

Figure 4.8: The working of the extinction vortex.

1. Habitat selection and evolutionary characteristics of the species making them rare: Species that are restricted to an uncommon habitat are rare because the habitats that support them are rare. Examples include species found in desert springs (dependent on a rare habitat; if there were too many springs it would not have been a desert!) and the pygmy hog-sucking louse restricted to the pygmy hog (its habitat itself is an endangered species!). Such species can be conserved only through protection and preservation of their habitat.

2. Limitations on the geographical range of the species: Species that cannot move to other habitats in their potential ranges are rare. Examples include those species that are found in a single lake, or trees that are restricted to a single island. Conservation of such species may make use of artificial transplantation to other suitable habitats.

3. Specific requirements and life history: Species that have low population densities, either because they have large-sized individuals that require more space, or because they are $K-selected$, meaning that they have evolved to prioritise parental investment over number of offsprings will be rare. In such cases, an integrated approach is needed for conservation, including not only the protection and preservation of their habitats, but also intensive *ex-situ conservation* [Chapter 10].

---

r/K selection

For a population, the logistic population growth is given as

$$\frac{dN}{dt} = rN \times \left(\frac{K-N}{K}\right)$$

where
N is the population size at time t,
K is the carrying capacity of the environment, and
r is the intrinsic growth rate

The $r-selected$ species are those that emphasise large growth rates (high r). They produce large numbers of offsprings that require little or no parental care. They typically occupy and exploit less-crowded ecological niches with low carrying capacities (low K), which means that individuals have a low probability of survival to adulthood. This strategy is useful because whenever the conditions become congenial, the fast rate of reproduction ensures that the ecological niche gets occupied. At the same time, because the conditions are unpredictable, there is little advantage in devoting large resources to the progeny generation by the parents. The typical characteristics of $r-selected$ species are

1. high fecundity, or the ability to produce an abundance of offspring,

2. early maturity onset,

3. short generation time,

4. small body size, and

5. the ability to disperse offsprings quickly and widely.

Examples of $r-selected$ species are rats, mice, bacteria and grasses. They produce large numbers of offsprings and provide little in the name of parental care.

The $K-selected$ species are those that occupy and exploit stable, highly crowded ecological niches near the carrying capacities (high K), which means that individuals have to *compete for resources*. Thus, parents need to invest large resources (providing food and protection in early growth stages, teaching the ability to hunt, etc.) in their offsprings to make them competitive. Because large resources need to be invested, only a small number of offsprings can be afforded. Thus, $K-selected$ species produce a small number of offsprings and display extensive parental care, giving individuals a high probability of survival to adulthood. The typical characteristics of $K-selected$ species are

1. low fecundity,

2. late maturity onset,

3. large generation time,

4. large body size,

5. large parental care, and

6. long life expectancy.

Examples of $K-selected$ species are tigers, elephants, dolphins and humans. They produce few offsprings in each generation and devote lots of time and resources in bringing them up.

Rarity may also be influenced by the impact of humans. The sensitivity of a species to human actions depends upon

1. Adaptability and resilience of the species: if these are low, a slight impact of human pressure will have large consequences for the species. This explains the loss of coral reefs which have low resilience.

2. Human attention: charismatic species like tigers are more sensitive because humans have high demand for their skin, bones and other body parts.

3. Ecological overlap between humans and the species: the greater the overlap, the greater the impact. This explains why grassland species are particularly sensitive — humans convert grasslands into agricultural fields.

4. Home range requirements of the species: species requiring larger home ranges are more sensitive to human impacts since they cannot survive in smaller areas. This explains why species like elephants become especially vulnerable. Elephants are large herbivorous animals (megaherbivores) and need a large home range to meet their dietary requirements without causing a complete ecological collapse through overgrazing. They are always on the move,

which permits the damaged vegetation to regenerate. If the habitat of elephants is reduced, they will be forced to consume most of the vegetation in a small area, which in the absence of sufficient time for regeneration will decimate many plant species. And then the elephants will die of hunger.

4.2.2 Red list classification

On the basis of vulnerability to extinction, the International Union for the Conservation of Nature (IUCN) classifies species in its "Red List" as follows:

1. Extinct (EX): A taxon is Extinct when there is no reasonable doubt that the last individual has died.

   e.g. dodo

2. Extinct in the Wild (EW): A taxon is Extinct in the Wild when it is known only to survive in cultivation, in captivity, or as a naturalised population well outside the past range.

   e.g. Northern white rhinoceros

3. Critically Endangered (CR): A taxon is Critically Endangered when available scientific evidence indicates that it is considered to be facing an extremely high risk of extinction in the wild.

   e.g. Javan rhinoceros

4. Endangered (EN): A taxon is Endangered when available scientific evidence indicates that it is considered to be facing a very high risk of extinction in the wild.

   e.g. tiger

5. Vulnerable (VU): A taxon is Vulnerable when the available scientific evidence indicates that it is considered to be facing a high risk of extinction in the wild.

   e.g. snow leopard

6. Near Threatened (NT): A taxon is Near Threatened when it has been assessed against the criteria and does not qualify for Critically Endangered, Endangered, or Vulnerable now, but is close to qualifying for, or is likely to qualify for, a threatened category in the near future.

   e.g. polar bear

7. Least Concern (LC): A taxon is Least Concern when it has been evaluated against the criteria and does not qualify for Critically Endangered, Endangered, Vulnerable, or Near Threatened.

   e.g. cow

8. Data Deficient (DD): A taxon is Data Deficient when there is inadequate information to make a direct, or indirect, assessment of its risk of extinction based on its distribution and/or population status.

   Many molluscs, fishes, and nocturnal birds and mammals have been evaluated, but could not be listed as Threatened because there was not enough information.

9. Not Evaluated (NE): A taxon is Not Evaluated when it is has not yet been assessed against the criteria.

   Most of the world's species, notably the invertebrates and other small life forms, fall into this category.

4.3 DESTRUCTION OF HABITAT

Destruction of habitats is a major reason for the threat to species. Destroy the habitat, and the resident species will face push factors everywhere. That is a sure recipe to push the species towards extinction. We can classify destruction of habitats into four categories:

1. Habitat degradation

2. Habitat fragmentation

3. Habitat displacement

4. Habitat loss

4.3.1 Habitat degradation

Habitat degradation is the process by which the quality of habitat for a given species gets diminished, and the habitat is unable to support the large number of individuals that it could do before being degraded. A good example is a polluted pond. Earlier, it could support a large number of fishes, but now with pollution, the number of fishes that can be supported has reduced. It still counts as a habitat since the fishes are still found in the pond. But it is a degraded habitat because the number of fishes has dropped.

Some common causal agents of habitat degradation are

1. contamination — due to pollution, including smoke, excessive fertilisers, pesticides and accumulative toxins,

2. trash, including ghost nets [Fig. 4.9a,b] and plastics [Fig. 4.9c,d,e],

3. soil erosion,

4. fires [Fig. 4.9f],

5. over-exploitation of water,

6. deforestation,

7. desertification due to over-grazing [Fig. 4.9g] and faulty cultivation practices,

8. draining, dredging and damming of water bodies,

9. over-exploitation of biota, and

10. introduction of exotic species and invasives [Fig. 4.9h].

4.3.2 Habitat fragmentation

Habitat fragmentation is said to occur when a natural landscape is broken up into small parcels of natural ecosystems isolated from one another, often in a matrix of lands dominated by human activities. It involves both loss and isolation of ecosystems. An example is seen in figure 4.10.

This is an issue because smaller fragments of habitat support far lesser number of species than their larger counterparts. This is because

1. Larger fragments have more diverse environments and thus more number of habitats able to support a larger biodiversity.

(a) An example of ghost net. Image source: NOAA [NOAA, 2020].

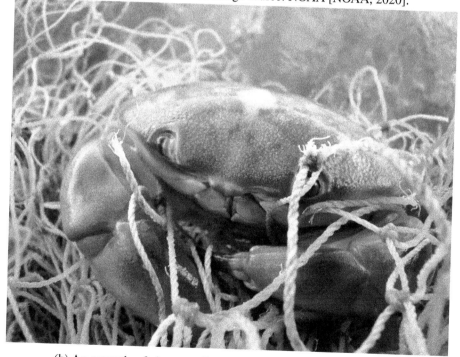

(b) An example of ghost net. Image source: NOAA [NOAA, 2020].

(c) Plastic waste near nesting sites of penguins at 'Boulders' Table Mountain National Park.

(d) Plastic waste in Mukurthi National Park where Nilgiri Tahr lives.

(e) Plastic waste in dung of rhinoceros at Manas Tiger Reserve.

(f) A site of forest fire in Kanha Tiger Reserve.

(g) Overgrazing is an important cause of desertification.

(h) A board describing invasive alien species Manitoka and Rooikrans that have invaded the fynbos vegetation in Table Mountain National Park of South Africa.

Figure 4.9: Examples of habitat degradation.

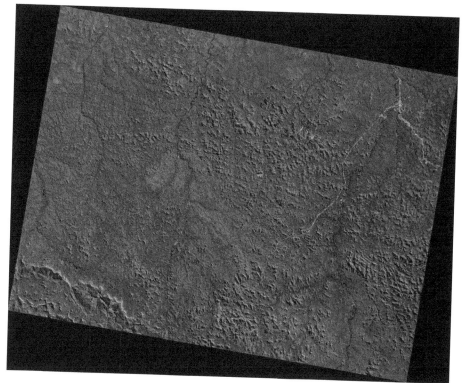

(a) The forest cover in 1975.

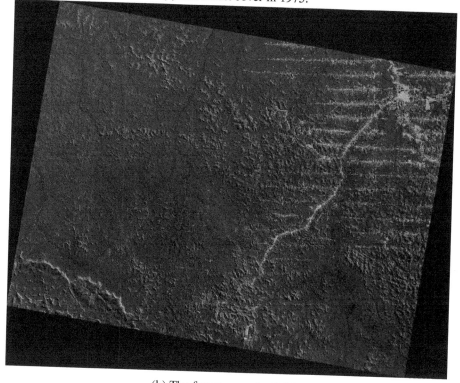

(b) The forest cover in 1986.

(c) The forest cover in 1992.

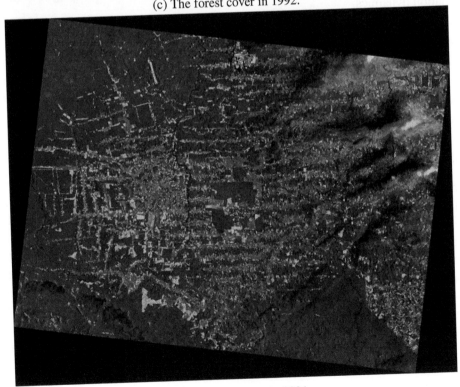

(d) The forest cover in 2001.

Figure 4.10: Example of habitat fragmentation and loss in Rondönia forests of Brazil [NASA, 2005].

2. Larger fragments are more likely to have both common and uncommon species; smaller fragments are more likely to have only common species.

3. Smaller fragments have smaller populations, so the chances of getting extinct are greater.

Thus habitat fragmentation — by reducing the size of habitat fragments — poses a threat to species. Some causal agents for habitat fragmentation are diversion of land for agriculture or other uses, and development of infrastructure, especially linear infrastructure [Section 4.5] that fragments the habitat.

The process of habitat fragmentation and loss occurs in several stages as detailed in figure 4.11.

4.3.3 Habitat displacement

Habitat displacement is the competitive exclusion of one species by another, often through displays of aggressive behaviour. It is commonly observed in birds, crabs, deer and other species, and is a natural phenomenon.

However, of late, habitat displacement is occurring on an unprecedented scale through the action of humans. Not only do humans displace other organisms to convert land into agricultural, infrastructural and for other uses, but they also displace the native fauna while grazing livestock. The pastoralists are often accompanied by dogs to display aggression and keep wildlife away.

When this happens, we can observe the shifting of wildlife to non-prime or sub-prime habitats such as hills or rocky patches. While these areas can provide shelter to wildlife, often they are resource-scarce in terms of food and water, and thus have a low carrying capacity. This plays a role in keeping the population of the wild animal at low levels, and may even bring it to the brink of small population dynamics.

4.3.4 Habitat loss

Habitat loss occurs when the quality of the habitat is so low that the habitat is no longer usable by a given species. This is the culmination of continued habitat degradation, fragmentation and displacement, but may also happen through catastrophic events such as large-scale fires, volcanic eruptions or wars. Some examples of lost habitats are shown in figure 4.12.

4.4 POPULATION VIABILITY ANALYSIS

Population viability analysis is an analysis of the viability — or continued survival — of a population. Population viability is defined as "the ability of a population to persist, or to avoid extinction."

In this method, the extinction probability of a single species population is assessed [Possingham et al., 2013] by integrating the data on its life history, demography and genetics, with the information on the variability of the environment, diseases, stochasticity, etc. — often utilising mathematical models and computer simulations — in order to predict whether the population will remain viable or go extinct in a decided time frame under various management options [Beissinger and McCullough, 2002].

Population viability analysis can be used to identify which threats are the most significant for a given situation. The method is discussed in detail in Chapter 8.

4.5 CASE STUDY: ECOLOGICAL IMPACTS OF LINEAR INFRASTRUCTURE

Improperly constructed linear infrastructure are some of the most glaring examples of habitat destruction by human beings. Linear infrastructure are defined as "those basic physical (and organisational) structures (and facilities) that are needed for the operation of a society or enterprise and

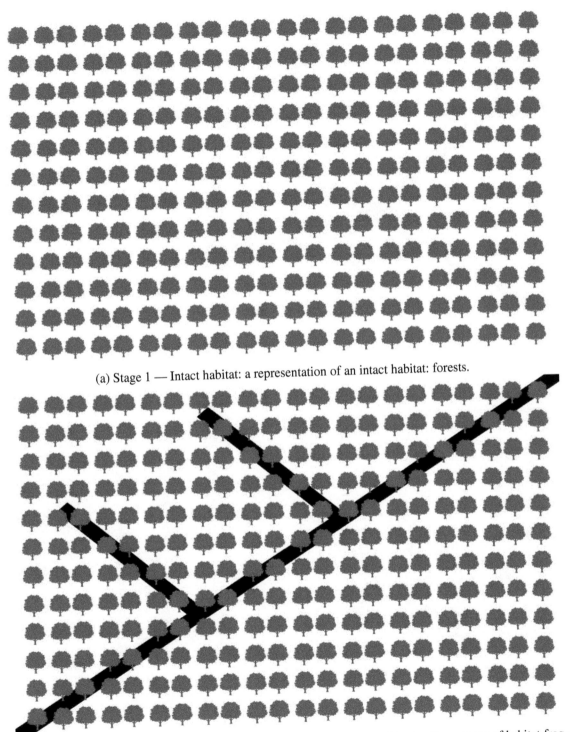

(a) Stage 1 — Intact habitat: a representation of an intact habitat: forests.

(b) Stage 2 — Dissection: construction of a road dissects the habitat and begins the process of habitat fragmentation.

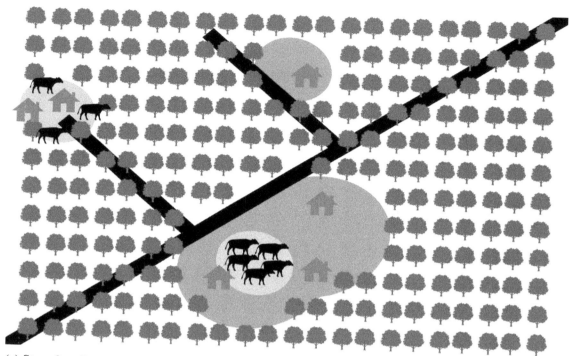

(c) Stage 3 — Beginning of perforation: small settlements emerge since humans are able to use the road to access previously inaccessible areas. The people and their animals start exploiting the forests, displacing and removing organisms.

(d) The presence of livestock in the forest is an indication of perforation.

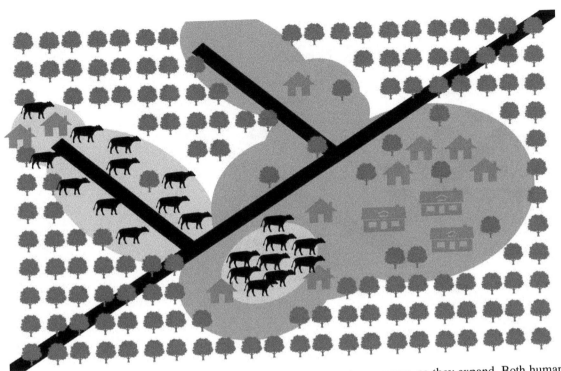

(e) Stage 4 — Increased perforation: the settlements have access to resources, so they expand. Both human and livestock population increases and more land is diverted for agriculture. Some permanent structures such as houses, school buildings, and hospital buildings get constructed.

(f) These forests in Mudumalai represent stage 4.

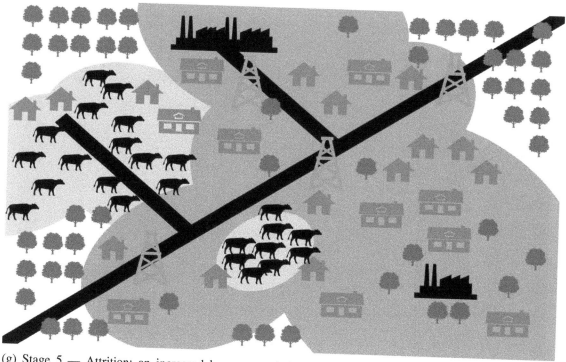

(g) Stage 5 — Attrition: an increased human population requires even larger resources like electricity, schools, hospitals and factories, which get provided. More people come into the region, and more forests get diverted, till most of the natural habitats have been lost.

(h) This image from Mudumalai represents stage 5.

(i) An extremely fragmented habitat in Masinagudi.

Figure 4.11: The process of fragmentation of habitat.

can be represented as straight or curved lines." In the context of ecological impacts, we are concerned with physical linear infrastructure [Fig. 4.13] such as roads, railways, power lines, canals, and pipelines.

Construction of linear infrastructure through wildlife areas can have large impacts on wildlife. This is because the animals that had been using the space before the linear infrastructure was built continue to use it. We can easily see several examples of wild animals crossing roads [Fig. 4.14a–c] or birds perching on power lines [Fig. 4.14d]. And this brings them on a collision course with humans.

The impact of linear infrastructure on wildlife can be studied under several heads:

1. Habitat destruction done during construction of the linear infrastructure: For constructing roads, railway tracks or canals, a large number of trees are felled and the ground is made level [Fig. 4.14e]. While levelling the ground, the earthwork often leaves large-sized holes on the ground at locations from which material was removed [Fig. 4.14f]. The felling of trees is a direct destruction of the wildlife habitat. At the same time, the holes in the ground pose threat to the animals for a very long time. (The animals break their bones and hurt themselves upon falling in these holes. In the rainy season, these holes get filled with water and pose the threat of drowning.) The melting of tar for black topping of roads not only creates pollution in the surroundings, but also increases the chances of forest fires.

2. Death of animals upon collision with vehicles: When animals get hit by vehicles moving at high speeds, it often leads to grave injuries and death [Fig. 4.14g–i]. Birds often miss the electricity wires on their flight paths and collide with the wires, often resulting in broken bones and death.

3. Barrier effect: At times when the traffic is very large, the animals may be unable to cross

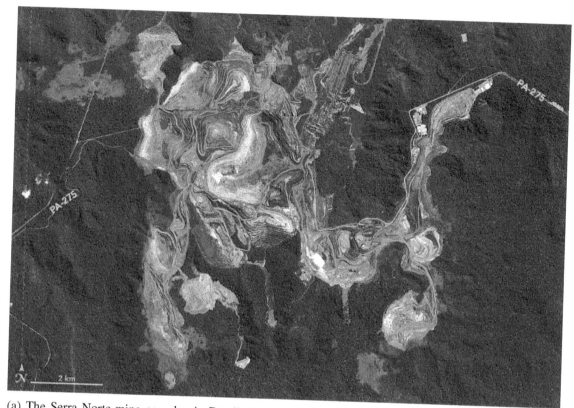

(a) The Serra Norte mine complex in Brazil showing extensive deforestation and loss of natural habitat. Image source: NASA [NASA, 2018b].

(b) Flooding due to Lower Sesan II dam resulted in large-scale loss of habitat in Cambodia in the year 2018. Image source: NASA [NASA, 2018a].

(c) The same region in Cambodia in the year 2017 for comparison. Image source: NASA [NASA, 2018a].

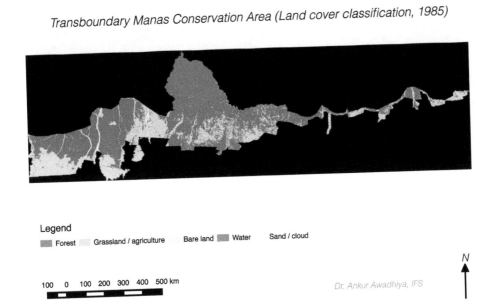

(d) The land cover classification of the Manas Trans Boundary Conservation Area in the year 1985.

Transboundary Manas Conservation Area (Land cover classification, 2003)

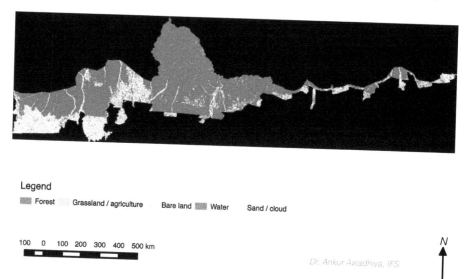

(e) The land cover classification of the Manas Trans Boundary Conservation Area in the year 2003. Note the conversion of grasslands into agricultural fields that show up as bare lands in off-agricultural season.

Transboundary Manas Conservation Area (Land cover classification, 2017)

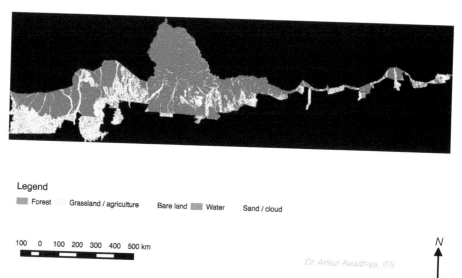

(f) The land cover classification of the Manas Trans Boundary Conservation Area in the year 2017. Note that the forests have not only reduced dramatically since 1985, but they have also become fragmented.

Figure 4.12: Examples of loss of habitat.

(a) Road.

(b) Railway line.

(c) Pipe line.

Figure 4.13: Examples of physical linear infrastructure.

the linear infrastructure such as road or railway track. This is known as "barrier effect." The effectiveness of the linear infrastructure in acting as a barrier depends upon several factors:

(a) traffic intensity: more volume of traffic effectively creates a 'wall' of vehicles that stops animals from crossing

(b) vehicle speed: higher speeds are more intimidating and increase the barrier effect

(c) driver sensitivity: use of headlight, horns, etc. can unnerve the animals

(d) presence and location of animal crossings: if suitable alternatives such as animal underpasses or overpasses are available, the animals will use them, and the road will not be a barrier to animal movement

(e) movement pattern of species: roads in an area where species move a lot will have a larger impact

(f) species specific preference of road use: some species may find roads more disconcerting than other species

(g) road edge features (e.g. height of embankment): very high embankments on the sides of roads increase the barrier effect

(h) time of day and year: the barrier effect is much more pronounced during the times of intense animal movement such as rainy season

(i) species diversity in the surroundings: the linear infrastructure's barrier effect impacts more species in more biodiverse areas

At low vehicle densities and speeds, the animals are able to cross the road. At very high vehicle densities and speeds, the barrier effect gets very pronounced and the animals are repelled by the road. It is at medium vehicle densities and speeds that the road becomes a death trap [Fig. 4.14j]. This is because while the vehicle densities and speeds are low enough to encourage the animal to cross the road, they are not low enough to permit a safe passage. The animals have a very high probability of getting hit by the fast-moving incoming vehicles.

4. Habitat fragmentation: Because of the barrier effect, many linear infrastructures lead to habitat fragmentation [Fig. 4.14k]. When animals are unable to traverse the landscape, their populations act as small, isolated populations and we start witnessing impacts of small population dynamics and the extinction vortex [Fig. 4.8].

5. Impacts of increased access: The increased access to the interiors of forests that becomes available to people has its own impacts. We can readily observe girdling and felling of trees along the roadside [Fig. 4.14l] done by opportunistic offenders and early settlers [Fig. 4.11c]. Increased access also increases littering [Fig. 4.14m] and poaching of animals in the forest areas.

6. Increased pollution: Vehicles cause air, sound and light pollution [Fig. 4.14n] which not only impacts animal health and causes degradation of habitat, but may also increase stress and cause changes in animal behaviour. Often the fecundity of animals reduces under stressful conditions, pushing the species towards small population dynamics.

7. Increased possibility of human-wildlife conflicts and resultant vindictive damage: When humans enter wildlife habitats, there is an increased chance of conflict situations, including injuries and loss of life and property [Fig. 4.14o,p]. One way in which people respond is by trying to reduce the wild animal numbers and densities in their surroundings through poisoning and poaching of wild animals. This is often facilitated by government agencies in the name of "making human settlements safer." This further threatens wildlife.

8. Changes in animal behaviour: Wild animals are generally apprehensive of human presence and try to avoid humans in natural situations. However, with increased interactions with humans, the fear of humans is gradually eroded. At times when people see a wild animal, they respond by offering food, either by compassion, due to religious reasons or for amusement. When wild animals are offered food again and again, they start treating the road as a means of getting food. This not only changes their behaviour [Fig. 4.14q–t] in profound ways, but also increases animal densities near the roads, enhancing the chances of the animals getting hit by vehicles or getting poached. The natural food of wild animals is typically low in salt, sugar and fats, but the human food is rich in these. Often we find animals not just habituated to human food, but also so dependent on it that they start showing aggression to get human food, further increasing the situations of conflict.

9. Increased spread of diseases: Increased interactions between wild animals and humans increase the chances of the spread of wildlife diseases to humans (zoonosis) and of human diseases to wildlife (reverse zoonosis). Zoonoses are often responded to by culling the wild animals to 'contain' diseases, while reverse zoonoses directly threaten wildlife.

10. Threats to animals by electricity infrastructures: Power lines cause electrocution of wild animals, especially perching birds and elephants that can reach the power lines [Fig. 4.14u]. During times of storms, the electricity wires may break, reach the ground, and electrocute other animals as well. Similarly, other infrastructures such as windmills [Fig. 4.14v] cause deaths of birds through impacts. The ground wires of electricity transmission lines are often thin and may not be readily visible to birds — which get impacted during flight. This is an important factor for the decimation of large-sized birds that have weak eyesights and are unable to manoeuvre swiftly on detecting the wires [Janss and Ferrer, 2000, Raab et al., 2011, Silva et al., 2014]. Common examples include bustards such as the Great Indian Bustard, a critically endangered species.

These negative impacts of linear infrastructure on wildlife can easily be overcome by taking suitable precautions. These are as simple as filling up the dug holes after construction, using bird diverters — small plastic pieces to signal birds about impeding collision — on electricity transmission wires, and provisioning of safe passages (like underpasses or overpasses) permitting animal movement. These measures are known as 'mitigation measures,' and often require nothing more than a sensitivity towards wildlife. It is truly a tragedy that these mitigation measures are not given the requisite attention in engineering institutions — to the detriment of wildlife conservation.

(a) A herd of elephants crossing a road in Kruger National Park.

(b) A wild boar crossing a road in Kruger National Park.

(c) A flock of birds crossing a road in Kruger National Park.

(d) A bird perching on a power line.

(e) Construction of roads is preceded by destruction and loss of habitats. Here we can observe clearing of forests and large-scale earthworks due to expansion of a road near Nauradehi Sanctuary.

(f) Large-sized holes left by earthwork for road construction in Nauradehi Sanctuary.

(g) A road kill (macaque) in Nauradehi Wildlife Sanctuary.

(h) A road kill (python) in Balaghat forest.

(i) A road kill (vervet monkey) in Kruger National Park.

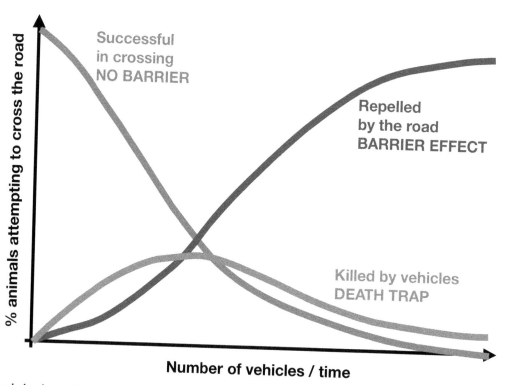

(j) The behaviour of roads changes from no barrier to a death trap and finally to an effective barrier for wildlife as the traffic volume increases.

(k) Linear infrastructure such as roads often fragment the habitats they traverse.

(l) The increased access to people that is provided by roads facilitates destruction of habitat. Here we can observe girdling and felling of trees near roadside.

(m) Roads increase access to forest areas, bringing wastes and garbage such as plastic bags that need to be regularly removed.

(n) Vehicles on road lead to air, sound and light pollution.

(o) Roads increase chances of human-wildlife conflicts by bringing humans and wildlife together. Here we can observe an elephant charging a vehicle in Kruger National Park.

(p) A vehicle damaged by collision with a nilgai in Nauradehi Sanctuary. The occupants of the vehicle were severely hurt and had to be hospitalised. The vehicle suffered major damages. The loss of life and property is often responded by retaliatory vengeful actions on wildlife by people, including poaching of animals.

(q) Increased access to people alters animal behaviour. A troop of langoors begging for food in Nauradehi Sanctuary.

(r) Bonnet macaque eating toast provided by people. At times, the wild animals become dependent on food provided by humans due to changes in food preference and reduced practice in gathering food themselves.

(s) Altered wildlife behaviour due to humans in Sariska Tiger Reserve.

(t) Macaque drinking cola in Mahabaleshwar forest.

(u) Power lines may electrocute perching birds.

(v) Birds routinely get hit by windmills such as these seen in Kukuru.

Figure 4.14: Impacts of linear infrastructure on wildlife.

CHAPTER 5

Wildlife monitoring

"What cannot be measured cannot be managed."

This statement lies at the heart of monitoring wildlife. To manage wildlife, we need to measure wildlife — have an idea about how many individuals are present, what their population characteristics are, how do they behave, and so on. Measurement of these attributes requires us to know the tricks of the trade — like how to identify species visually, or from their vocalisation, or from the pug marks visible in the soil, how to count individuals using photographs from a camera trap, or by doing some measurement, and how to read the actions and behaviours of individuals to get a grasp on their health status. In this way, we can define wildlife monitoring as "the science of keeping track of animal characteristics — their movement patterns, habitat utilisation, population demographics, poaching incidents, disease breakouts, etc." In this chapter, we explore some techniques of monitoring wildlife.

5.1 COUNTING WILDLIFE: ACCURACY, PRECISION & BIAS

When we need animal numbers, the first question to consider is the *precision* of the count. Precision is how close the measured values are to each other. Suppose we make five measurements of animal density, and obtain the values: 101, 103, 102.5, 101 and 102 animals per square kilometre. Since these values are close to each other, we'll call the measurements precise. On the other hand, if the values obtained are: 101, 130, 210, 94 and 50 animals per square kilometre, the values would be said to be less precise. Deciding on the level of precision is important since it is rather simple to make a broad estimate of the animal numbers. However, the effort required to get to the precise figures is relatively enormous — especially since multiple factors that can affect the readings have to be kept under control. Since the resources — in terms of time, money, manpower, equipment, etc. — are limited, we may be obligated to avoid the precise count in several instances.

The second question to consider is the *accuracy* of the count. Suppose there are 477 individuals, but when we do count, we arrive at the figure of 474. Not bad, right? It's pretty close. But suppose we count 414 individuals, then? Would it be *okay* to report this figure? What about 244? Where do we draw a line?

Accuracy is how close the measured values are to the correct value. Suppose the correct density of animals is 103 animals per square kilometre. If our obtained values are: 101, 103, 102.5, 101 and 102 animals per square kilometre, we'll call our readings accurate, since they hover around the correct value of 103. On the other hand, if the values are: 151, 153, 152.5, 151 and 152, then although the values are close to each other (precise), they are not close to the correct value (103). Thus, in this case, the readings will not be called accurate.

Ideally we wish to have results that are both accurate and precise [Fig 5.1a]. We define bias in the measurement [Fig 5.1b] as the difference between the mean of the measured values, and the reference value. If the reference value is the true value, then bias indicates the error in the measurement, or the accuracy of the measurement.

To remove bias, we need to *calibrate* the instrument or the method of measurement. Thus, in our example [Fig 5.1a], the gun of shooter (c) may have a bend towards the right and upwards. When this bend is corrected for — i.e. the gun is calibrated — the shots will start falling on the bull's eye, leading to accuracy.

5.2   CENSUS VS. SAMPLING

Counting of wildlife is done in two ways — census and sampling. *Census* is a procedure of systematically recording information about *all* the members of a population. For example, in the case of a tiger census, we aim to take photographs of *all* the tigers in a population, identify each of them using their stripe patterns, and thus record the information about *each* tiger. This requires deployment of large amount of resources in the form of large number of devices, many personnel, time and money — to ensure that all the tigers are accounted for, and no tiger is missed. Since we often have a dearth of resources, census is often done only periodically, and that too for those animals that are of critical importance. You will not hear about a census of rabbits or deer, except in some very specialised circumstances.

For the majority of animals, we employ sampling. *Sampling* is a procedure in which information about just a small fraction of the population is recorded, and this reading is then *extrapolated* to the complete population. Thus we may measure animal density at one, or a few representative location(s) and say that we can use this (these) reading (readings) as a representation of the average animal density for the complete population. Multiplying the average density with the area where the population is found will give us an *estimate* of the population count.

When doing sampling, the most important objective is to secure the *representative* sample — one which can reproduce the important characteristics of the whole population as closely as possible. For instance, we will not measure the density in *outlier* areas — those with very high or very low densities in comparison to the target population — but will measure density at those location(s) that represent the *average* population density of the target population. The crux of sampling is to find this *representative sample*.

To do this, we begin by defining the population. Population is "the aggregate of units from which a sample is chosen." Thus all the tigers living in a landscape under study will form a population, from which a representative sample needs to be drawn (chosen).

Next we define sampling units — administrative, natural or artificial units that are well defined, identifiable in the forest, and on which observations can be made. Thus we may divide the forest into grids of 10 km × 10 km, and call each square as a sampling unit. Or we may make use of well-defined administrative or management boundaries such as forest compartments — and use them as sampling units. We prepare a list of these sampling units — say by giving each sampling unit a numerical (such as compartment number 51) or an alphanumeric identity (such as grid N31E54). This list of sampling units is called a 'sampling frame.' Out of this list, a sample will be drawn. The sample will comprise of one or more sampling units in the sampling frame that are selected according to *some specified procedure* — say, by taking every nth sampling unit in the sampling frame.

The value of *n* is provided by the *sampling intensity* — the ratio of the number of units in the sample to the number of units in the population. A high sampling intensity means that a large number of units in the population are part of the sample. This requires a large provisioning of resources for the sampling exercise. A low sampling intensity means that a small number of units in the population are part of the sample. This requires a small provisioning of resources for the sampling exercise.

As the sampling intensity increases, more and more number of units in the population form part

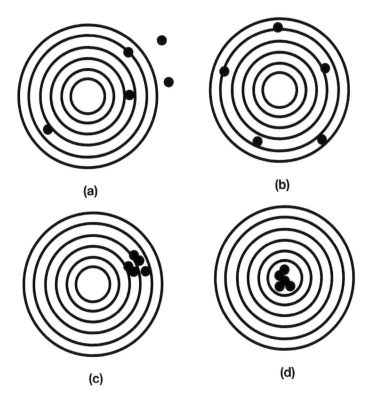

(a) Difference between precision and accuracy, exemplified by the shots of different shooters on a target board. The first shooter is neither precise nor accurate. The second shooter is accurate but not precise. The third shooter is precise but not accurate. The fourth shooter is both precise and accurate.

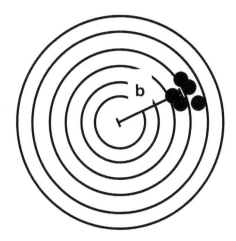

(b) Bias is the difference between the mean of the measured values, and the reference value — here, the bull's eye.

Figure 5.1: Precision, accuracy and bias.

of the sample. This increases the representativeness of the sample to the population, and results in an increased accuracy of measurement, albeit at the cost of large provisioning of resources for the sampling exercise. However, the marginal benefit of increasing sampling intensity — i.e. the increase in accuracy achieved through each additional unit of effort — goes on decreasing. After all, once the sampling intensity is enough to result in a representative sample, more sampling will not increase the accuracy of the measurement, but will only increase the effort of doing the sampling exercise. Thus, we often reach a 'sweet spot' of sampling intensity in the middle — and prefer sampling intensity that is neither too low (which would result in low accuracy of measurement), nor too high (which would require resources and effort disproportionate to the increase in accuracy).

Sampling may be categorised on the basis of the 'specified procedures' used to select the sample from the sampling frame, giving various kinds of sampling:

1. Simple random sampling: In this procedure, each possible combination of sampling units has the same chance of being selected into the sample. This is often achieved using selection by lottery, or rolling of dice, or by using a list of random numbers.

2. Systematic sampling: In this procedure, every kth unit is selected starting with a number chosen at random from 1 to k as the random start. For example, we may say that every 3rd unit in the sampling frame will be selected, beginning with the 2nd unit in the sampling frame. In this case, the sample shall comprise the following units:

$$2, 5, 8, 11, 14, 17, \ldots$$

3. Stratified sampling: In this procedure, we divide the population into *strata* — internally homogeneous layers for which precise estimates can be made using a small sample. For instance, we may say that hills form a stratum, plain woodlands form a stratum, and plain grasslands form a stratum. In this case, while the density of animals in the hills may be very different from the density of animals in the plain woodlands, we make use of the fact that the density of animals in one part of the plain woodland will not be *very* different from the density of animals in another part of the plain woodland. In other words, while the population is heterogeneous, each of the strata are internally homogeneous, and so densities in each stratum can be *precisely* and *accurately* measured using a few measurements. Suppose we get the densities as ρ_{hill}, $\rho_{plain\ woodland}$ and $\rho_{plain\ grassland}$. Then the estimate of animal count is given as

$$\text{Estimate} = \rho_{hill} \times \text{Area}_{hill} + \rho_{plain\ woodland} \times \text{Area}_{plain\ woodland} \\ + \rho_{plain\ grassland} \times \text{Area}_{plain\ grassland}$$

Since the individual densities of each stratum and area of each stratum can be precisely and accurately measured due to the homogeneity of the stratum, the final estimate obtained through stratified sampling is often much more precise and accurate than a measurement made using simple random or systematic sampling.

4. Multistage sampling: In this procedure, large sized units are selected and then a specified number of sub-units from each selected large units is chosen. For instance, we may select ten states and choose five districts in each selected state for measurements. This often results in a more representative sample than that obtained through simple random sampling. In simple random sampling, it is possible that all the selected units come from one state or area — and thus be less representative of the population. Multistage sampling ensures that the samples obtained come from multiple states or areas to increase the representativeness of the sample to the population.

We now look at two examples of computation of animal density.

Example 1

A park manager conducts a population estimation exercise within a protected area. He samples 18 quadrats with line transects and obtains the density estimates for sambar deer (*Rusa unicolor*) as shown in table 5.1.

Table 5.1: Sambar densities in different beats.

Beat number	Sambar density (number per sq. km)
1	8
2	5
3	6
4	5
5	8
6	3
7	7
8	7
9	6
10	2
11	8
12	9
13	4
14	7
15	8
16	2
17	1
18	5

We assume that each beat has the same area. In this case, the average density will be calculated as:

$$\text{Average density} = \frac{\Sigma(\text{Density in the beat})}{\text{Total number of beats}}$$

$$\implies \text{Average density} = \frac{8+5+6+...+5}{18}$$

$$\implies \text{Average density} = \frac{101}{18}$$

$$\implies \text{Average density} = 5.61 \text{ animals per square kilometre}$$

Example 2

The groups sizes of chital (*Axis axis*) in the core and buffer zones of Corbett Tiger Reserve, Uttarakhand were recorded during winter 2009 (data given in table 5.2). We need to estimate the mean group size, standard deviation, standard error, range and coefficient of variation, and comment on the results obtained.

The mean group size in the core zone is calculated as:

$$\text{Mean group size} = \frac{\Sigma(\text{Group sizes in core})}{\text{Total number of groups}}$$

Table 5.2: Data for group sizes of chital in core and buffer of Corbett Tiger Reserve in Winter 2009.

Core zone	Buffer zone
26	26
24	11
25	07
27	03
23	15
25	19
27	22
26	30
24	34
25	40

$$\implies \text{Mean group size} = \frac{26 + 24 + 25 + \ldots + 24 + 25}{10}$$

$$\implies \text{Mean group size} = \frac{252}{10}$$

$$\implies \text{Mean group size} = 25.2 \text{ animals per group}$$

The mean group size in the buffer zone is calculated as:

$$\text{Mean group size} = \frac{\Sigma(\text{Group sizes in buffer})}{\text{Total number of groups}}$$

$$\implies \text{Mean group size} = \frac{26 + 11 + 7 + \ldots + 34 + 40}{10}$$

$$\implies \text{Mean group size} = \frac{207}{10}$$

$$\implies \text{Mean group size} = 20.7 \text{ animals per group}$$

The standard deviation for mean group size in the core zone is calculated as:

$$\text{Standard deviation (population)} = \sqrt{\frac{\Sigma(x - \mu)^2}{N}}$$

$$\implies \text{Standard deviation (population)} =$$

$$\sqrt{\frac{\Sigma[(26 - 25.2)^2 + (24 - 25.2)^2 + \ldots + (25 - 25.2)^2]}{10}}$$

$$\implies \text{Standard deviation (population)} = \sqrt{\frac{15.6}{10}}$$

$$\implies \text{Standard deviation (population)} = \sqrt{1.56}$$

$\implies$ Standard deviation (population) = 1.249 animals per group

The standard deviation for mean group size in the buffer zone is calculated as:

$$\text{Standard deviation (population)} = \sqrt{\frac{\Sigma(x-\mu)^2}{N}}$$

$\implies$ Standard deviation (population) =

$$\sqrt{\frac{\Sigma[(26-20.7)^2+(11-20.7)^2+...+(7-20.7)^2]}{10}}$$

$\implies$ Standard deviation (population) = $\sqrt{\dfrac{1296.1}{10}}$

$\implies$ Standard deviation (population) = $\sqrt{129.61}$

$\implies$ Standard deviation (population) = 11.385 animals per group

The standard error for mean group size in the core zone is calculated as:

$$\text{Standard error} = \frac{\sigma}{\sqrt{n}}$$

$\implies$ Standard error = $\dfrac{1.249}{\sqrt{10}}$

$\implies$ Standard error = 0.395 animals per group

The standard error for mean group size in the buffer zone is calculated as:

$$\text{Standard error} = \frac{\sigma}{\sqrt{n}}$$

$\implies$ Standard error = $\dfrac{11.385}{\sqrt{10}}$

$\implies$ Standard error = 3.600 animals per group

The range for group size in the core zone is calculated as:

$$\text{Range} = \text{Highest value} - \text{lowest value}$$

$\implies$ Range = 27 − 23

$\implies$ Range = 4 animals per group

The range for group size in the buffer zone is calculated as:

$$\text{Range} = \text{Highest value} - \text{lowest value}$$

$\implies$ Range = 40 − 3

$$\implies \text{Range} = 37 \text{ animals per group}$$

The coefficient of variation for group size in the core zone is calculated as:

$$CV = \frac{\sigma}{\mu} \times 100\%$$

$$\implies CV = \frac{1.249}{25.2} \times 100\%$$

$$\implies CV = 4.956\%$$

The coefficient of variation for group size in the buffer zone is calculated as:

$$CV = \frac{\sigma}{\mu} \times 100\%$$

$$\implies CV = \frac{11.385}{20.7} \times 100\%$$

$$\implies CV = 54.998\%$$

Comment on the results obtained

The group sizes of chital in the core zone are more or less similar, as shown by a small range value of 4 and a coefficient of variation of 4.956%. However, the group sizes of chital in the buffer zone are extremely variable, as shown by a larger range value of 37 and a coefficient of variation of 54.998%. The coefficients of variation suggest that the standard deviation is very far from the mean value in the case of the chital groups in the buffer zone, while the standard deviation is close to the mean in the case of the chital groups in the core zone.

These numbers also provide an indication of the habitats in the core and buffer zones. Since core zones are mostly unfragmented and uniform, the group sizes of chital groups show little variation from one group to the next. On the other hand, since the buffer zones are relatively fragmented and non-uniform, often with high anthropogenic influences, each chital group in the buffer zone will show a difference from other groups, depending on the patch of habitat that was available to it.

In this way, we may utilise statistical information to make sense of, or even to predict, the ecological characteristics of wildlife habitats.

5.3   DISTANCE SAMPLING

When we do sampling to find the density of animals, we select representative locations, count the animals and divide the count by the area to get an estimate of the density of animals at the chosen site. The estimates of density at several sites is averaged to get the average density of animals in the population.

When we count the number of animals in a sample site, can we be 100% sure that we have counted all the animals? Probably not. There are a number of factors through which we may miss the animals. Wildlife habitats often have trees, shrubs, grasses and other vegetation, and the animals may hide behind these vegetation. Even otherwise, both predators and preys have body colours that camouflage them against the background so that they may avoid detection. This camouflage may prevent us from counting some animals that are present in the sample site. Some animals also use behavioural techniques such as freezing (becoming motionless) when they sense danger. When the animals hold their movement, it may be difficult for us to detect them. Since these factors are invariably present in any wildlife setting, we may agree that the count of animals in the sample site is often an underestimate. This leads to the final count also being an underestimate. We require alternative procedures to solve this issue of non-detectability of animals.

Distance sampling is a sampling method for computing animal densities *taking into account the non-detectability of animals*. It is thus an advancement over 'normal' sampling.

To illustrate, let us consider the use of strip plots. Suppose we take k strips each of length l and half-width w.

Then, area of one strip plot will be $2w \times l$, and the total area of all the strip plots will be

$$a = k \times 2w \times l$$

If n animals are counted in this area a, then the density of animals in the sample plots can be given as:

$$d = \frac{n}{a}$$

We estimate that the density of animals in our region of interest is the same as the density of animals discerned from the sample plots:

$$\hat{D} = d = \frac{n}{a}$$

In this case, the estimate of total animals in the area of interest A will be:

$$\hat{N} = \hat{D} \times A = \frac{n}{k \times 2w \times l} \times A = \frac{nA}{2wL}$$

where L is the total length traversed, also called effort:

$$\text{Effort, } L = k \times l$$

You'll notice that we use hats in $\hat{N}$ and $\hat{D}$. This is to differentiate the quantities that are *estimated* from quantities that are *measured* or *counted*. The estimated quantities are given in 'hats.'

Thus, while N represents the actual count of animals as found in a census, $\hat{N}$ represents the total count of animals as estimated using the sampling exercise.

Similarly, while D would represent the actual density of animals as found using a census count with the total area of the region of interest, $\hat{D}$ represents the estimate of density of animals as found through a sampling exercise.

While N and D are the accurate values, $\hat{N}$ and $\hat{D}$, being estimates, may or may not be accurate.

5.3.1 How is distance sampling different from plot sampling?

Distance sampling aims to accommodate for the fact that even n and d have errors — they are, at best, mere *estimates*. This is because plot sampling assumes that all the animals in the sample plots are detected and counted, which is often an incorrect assumption, as shown in figure 5.2.

Distance sampling takes this factor of non-detection into account by basing itself on the premise that not all animals in the sample site are detected. It tries to mathematically estimate the number of animals that were missed in the counting exercise to arrive at a better estimate of animal count or density.

The corrected estimate is given as:

$$\hat{n} = \frac{n}{\hat{p}}$$

where
$\hat{n}$ is the corrected estimate of animal numbers,

(a) A hyena partly hidden in grasses in Kruger National Park. Such animals may be missed during 'counts.'

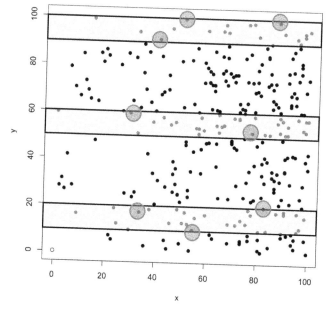

(b) Many animals, even those that are close to the point of observation, may at times get missed for several reasons. The undetected animals in these strip plots are represented by red circles.

(c) Animals may be hidden in plain sight. This is a tree in Kruger National Park with the animal barely visible.

(d) This is a close-up image of the tree in (c). The leopard is now clearly visible.

(e) This is a patch of *Lantana camara* in Nauradehi Wildlife Sanctuary with the animal barely visible.

(f) A careful observation of the patch in (e) reveals a tiger hiding inside.

Figure 5.2: The question of detectability of targets.

n is the uncorrected estimate of animal numbers as found in the sample plot by considering that all the animals in the plot have been detected, and

$\hat{p}$ is the estimate of the probability of detection of animals.

For instance, suppose $\hat{p}$ is estimated to be 0.75. This would mean that in any sampling, only 3 out of 4 animals will probably get detected.

Thus, if 30 animals were detected in plot sampling, there were actually 40 animals in the sample plot, considering the estimated probability of detection. We can write this as:

$$\hat{n} = \frac{n}{\hat{p}} = \frac{30}{0.75} = 40$$

By acknowledging the fact that not all animals are detected in sampling, and that adjustments have to be made using the probability of detection, we get another prominent distinction between plot sampling and distance sampling — the width of the strips. Since plot sampling assumes that all the animals inside the strip get detected and counted, the width of the strips must invariably be small so that this assumption remains valid. This is because the probability of detection of animals goes on decreasing with the distance from the transect line. For the plot sampling assumption of full detectability to hold true, the distance from the transect line, therefore, cannot be large.

On the other hand, since distance sampling assumes that not all animals will be detected, the strips can be much wider. For every width of the strip, there will be some $\hat{p}$, which can be estimated.

5.3.2   What does $\hat{p}$ depend on?

Under field conditions, the estimate of the probability of detection of animals — $\hat{p}$ is often an extremely tricky quantity to estimate, since it is affected by several factors, many of which may not even be under our control. Some factors influencing $\hat{p}$ are listed below:

1. **The characteristics of the terrain**: Flat, clear terrains provide little scope for animals to hide while providing the observer a large field of view. On the other hand, animals may not be seen in tall grasslands, thick bushes or undulating terrains. Thus, the terrain has a large bearing on the value of $\hat{p}$.

2. **The nature of the sampling exercise**: Sampling using vehicles (e.g. vehicular transects) may result in larger number of detections than sampling through walking on the floor (such as walking transects), not only because they provide a higher vantage point of observation, but also because the animals, habituated to vehicles by their park experiences, often feel more comfortable and do not run away.

3. **The nature of the transect**: Transects on areas that are more frequented by animals will have larger detection probabilities. Examples include some trails and roads.

4. **Local traditions**: In areas where locals feed animals, the animals might come closer to the observer, facilitating their detection.

5. **The characteristics of the animals**: Some animals are bold — they approach the observer. Some animals are shy — they avoid the observer. Some animals may have a keen sense of sight, sound or smell; others may not. The characteristics of the animals under observation have a large bearing on the value of $\hat{p}$.

6. **The colour of the dress worn by the observer**: Bright colours will make the observer easily detectable by the animals. While some shy animals might move away (facilitating their detection due to movement) or freeze (hampering their detection due to absence of movement), some other bold animals might approach the observer (facilitating their detection).

7. **The perfume worn by the observer**: Since many animals have a keen sense of smell, perfume will affect the detectability of the observer, resulting in similar effects as listed above.

8. **The food eaten by the observer**: Many foods leave a trace scent, which might be sensed by the animals, affecting $\hat{p}$.

9. **The mental state and fatigue of the observer**: Under conditions of mental stress or fatigue, an observer might not take note of all the animals, affecting the estimate of detectability of the animals.

10. **The size of the observer group**: When there are more observers, there are more number of eyes that could facilitate the detection of animals. On the other hand, if the group members talk, the sound could signal the animals, affecting their detectability, besides hampering the concentration of the observers.

11. **The direction of wind**: When wind blows from the observer(s) towards the animals, human scent quickly reaches the animals. On the other hand, when wind blows from the animals towards the observer(s), then the human scent does not reach the animals that fast. Animals react to scent; hence the direction of wind may also affect the $\hat{p}$.

12. **The weather on the day of observation**: Overcast skies will give very different results from clear skies, since weather impacts the activity of animals. When animals are more active and keep moving, they are more easily detected.

Since there are so many factors affecting $\hat{p}$, the value of $\hat{p}$ is often estimated using computers to process large amounts of field data. For mathematical derivations and examples of how this is done, the reader may refer to author's book "Fundamentals of Distance Sampling" published by the Wildlife Institute of India.

5.4 SIGN SURVEYING

It is clear that counting wild animals, nay, even *detecting* wild animals is often quite difficult. The animal may be missed even when we are right next to it. In such a scenario, we need to explore options — other than direct sighting — to monitor the animals. One such option is sign surveying.

The method of sign surveying is based on the fact that if an animal exists in an area, it will leave certain signs that can be used to identify the animal. To elucidate, consider the foot impressions that animals create when they walk. Animals will often go to the water bodies to drink water [Fig. 5.3a], and a large number of animal tracks can be observed on the soft mud. Bare paths on the ground are often covered with fine dust, and here again we may observe foot impressions. We can easily differentiate between a mark created by a tiger (called a pugmark, as observed in Fig. 5.3b) and a mark created by an elephant [Fig 5.3c] — both have very distinct characteristics. A tiger pugmark is smaller than an elephant track, and clearly reveals the digits on the tiger's paw. An elephant's track is larger and circular-shaped. Similarly, the hoof marks of herbivores such as sambars [Fig. 5.3d] will bear impressions of the hooves — tips of the toe.

The excreta of an animal is also an excellent sign that can be used to identify the animal. The scats of carnivores [Fig. 5.3e] often have hairs and bones of their prey, while the dung of herbivores [Fig. 5.3f] often has undigested plant fragments. The size, shape, colour and consistency of the excreta can be used to identify the animal. For a more accurate identification, the cells found in the excreta can be put through DNA analysis.

Territorial animals — those that defend their territories — often leave signs to announce their territory to other animals. Territoriality is easily observed in carnivores such as tigers and lions, but herbivores such as sambars may also show territoriality — especially during the mating season.

The territory markings left by the animals can also be used to identify the animals. These territory markings may be in the form of scratch marks on trees [Fig. 5.3g], scrape marks on the ground and strong-smelling urine or other secretions.

It is also possible to infer the species by looking at the feeding site. The kill of a tiger [Fig. 5.3h] will show a very clean opening of the prey — often from the rump. The kill of wild dogs will show multiple bite marks from the members of the pack. Even many herbivores can be identified — from the dig holes that they create to reach roots and tubers [Fig. 5.3i]. Similarly, birds can be identified from the shape, size and construction of their nests [Fig. 5.3j] and the shape, size and colour of their eggs [Fig. 5.3k].

Other signs include

- animal caches: e.g. squirrels that hoard nuts,

- drag marks: when carnivores drag the carcass of their prey, it creates drag marks which often can provide information about the size of the carnivore,

- scrapes on ground, especially defecation scrapes: these are often created by members of the dog family — they throw dirt to cover their faeces,

- animal calls, and

- remains of animals, especially antlers and bones.

Good locations to search for animal signs include

- forest paths frequented by animals — especially those that are covered with dust on which animal tracks can be readily observed,

- dusty and damp grounds,

- river beds,

- known animal trails,

- around water holes, and

- near natural salt licks.

After recording animal signs, we may convert the data into estimates of animal occupancy or density. Thus we may plot the locations where tigers roam — inferred by the presence of their signs. This gives an estimate of animal occupancy — the percentage of locations that are occupied by tigers. We may also estimate the density of animals by comparing the number of signs found per unit area (square kilometre) or per unit effort of finding signs (such as number of animal signs found per hour of surveying). The number of animals can be estimated using the amount of excreta found per unit sampled area, the animal's daily rate of defecation and the decay characteristics of excreta — time the excreta takes to decay in the sampled environment.

Thus, sign surveying can be used to obtain data to monitor wild animals even without actually observing them.

5.5 RADIO-TELEMETRY

Radio-telemetry is a technologically sophisticated technique to monitor wild animals. The word *telemetry* is derived from the roots *tele* — meaning distant, and *metron* — meaning measurement. Thus *telemetry* is the process of measuring from a distance, or transmitting and recording the

(a) Water holes, such as this in Timli Forest Range, are good places to look for animal signs.

(b) The pugmark of a tiger observed in Sariska Tiger Reserve.

(c) An elephant track observed in Timli Forest Range.

(d) A sambar hoof mark observed in Timli Forest Range.

(e) A scat observed in Timli Forest Range.

(f) The dung of wild ass observed in Wild Ass Sanctuary.

(g) A scratch mark observed in Timli Forest Range.

(h) The kill of a tiger observed in Balaghat Forest Division.

(i) A dig hole observed in Timli Forest Range.

(j) Nests of weaver birds observed in Panna Tiger Reserve.

(k) A nest of ground-dwelling bird observed in Wild Ass Sanctuary.

Figure 5.3: Some examples of animal signs.

readings of an instrument (measurement) from a distant location. Radio-telemetry is telemetry done using radio waves.

In the context of wildlife management, radio-telemetry refers to the tracking of wild animals through the use of sensors, transmitters and receivers. The most common format is VHF radio tracking using very high frequency radio waves. In it, the telecommunication consists of three parts:

1. Transmitter

2. Receiving antenna

3. Receiver

The transmitter is fitted to the animal through a collar [Fig. 5.4a] and transmits radio signals in the very high frequency spectrum. It consists of four components:

1. a power source (battery),

2. an electronics package (circuit board and crystal oscillator),

3. a transmitting antenna, and

4. an attachment method (collar).

To attach the collar to the wild animal, the animal often needs to be immobilised and captured. The most important considerations are the weight of the transmitter (it should not exceed 5% of the body weight of the animal), the comfort of the animal (the transmitter should not obstruct the movement and activities of the animal) and the lifetime of the transmitter (continuous transmission reduces the life of the battery, meaning that the transmitter will function for a smaller period of time).

The receiving antenna may use a single antenna or multiple antennae. Yagi antennae are preferred due to their robust design and directionality [Fig. 5.4b]. The antennae are connected to the receiver through a coaxial cable. The receiving antenna is held up and moved in different directions. Being a directional antenna, it will pick up strong transmission in the direction where the transmitter is, and weak transmission in other directions. The antenna may also be used in an I fashion [Fig. 5.4c] for improved directionality, albeit at the cost of reduced sensitivity.

Receivers are devices that help isolate the frequency of the transmitter and depict it in the form of an audible pulse or the movement of a needle [Fig. 5.4d]. They often deploy digital signal processing and filters to improve the signal: noise ratio, and may also have a data logging unit.

To determine the position of the collared animal using radio-telemetry, two methods can be deployed:

1. Triangulation [Fig. 5.4e]: In this method, the operator stands at a certain location and moves the antenna to the direction where the signal is the strongest. The directional bearing is noted using a compass. Then the operator moves to a second location and repeats the process. With the two locations defining the base vertices of the triangle with the side length of d, and the two angles α and β discerned from the compass bearings, the third vertex can be computed. This vertex is the position of the collared animal.

   In practice, multiple locations and multiple bearings are used to reduce the error in positioning of the animal.

2. Homing [Fig. 5.4f]: In this method, the operator stands at a certain position and moves the antenna to the direction where the signal is the strongest. Then keeping the direction fixed, he/she starts moving towards the animal. As the distance to the animal reduces, the signal intensity increases and louder beeps are heard. These are then compensated by reducing the gain in the receiver. At very close distances, the antenna may be deployed in the shape of alphabet I, or the antenna may be removed and just the coaxial connection used as an antenna. In this manner, the operator is able to reach the animal through homing.

Advancements in technology have helped create small-sized transmitters with long battery life span, permitting deployment to several small-sized animals. Modern devices also permit automatic recording of coordinates of the animal using GPS (Global Positioning System) at set intervals, together with data about the direction of movement, the speed of the animal, the altitude/depth of the animal and the health and physiological indicators of the animal. All these data can also be retrieved through satellites without the need to re-capture the animal to retrieve the storage card. Such advancements are enhancing the ease of field deployment of these tracking devices and permitting automated fine-scale monitoring of numerous wild animals.

5.6 BEHAVIOURAL MONITORING

Behaviours are the ways in which organisms respond to each other and to particular cues in the environment. Examples include basking [Fig. 5.5a,b] wherein animals — especially cold-blooded animals such as crocodiles, snakes and turtles — respond to the cue of warmth of the Sun, especially in cold days. Being cold-blooded organisms, they need to warm their bodies for optimal functioning. Using an external source of heat — such as the Sun — is an energy-conserving strategy. We can easily observe these animals basking in the Sun in the early mornings of winter season.

We may also easily observe foraging and feeding behaviours [Fig. 5.5c–f]. Herbivorous feeding behaviours involve reaching a location of food and eating the food. Depending on the diet and the availability of food, these may require selecting the best time to have the fruits — when they are

(a) A transmitter being worn by a tiger in Panna Tiger Reserve. The transmitter emanates radio signals at a fixed frequency.

(b) A receiving antenna in the normal mode of operation. Directional antennae are used to discern the direction of the animal wearing the transmitter.

(c) At closer distances, the antenna is deployed in I-fashion for improved directionality.

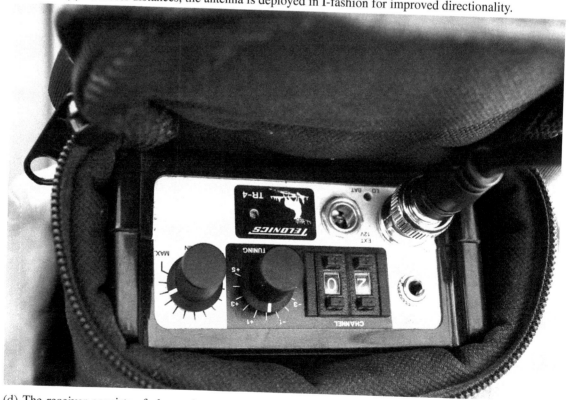

(d) The receiver consists of electronic components for selection of frequency and conversion of received radio signals into sound or other signatures.

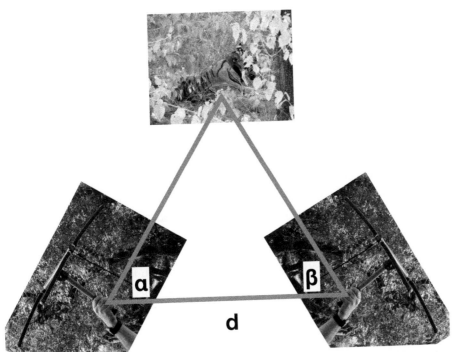

(e) In triangulation, the direction to the transmitter is noted from two locations. These two locations form the base of a triangle, and the directions permit calculation of the third vertex, which is the location of the transmitter.

(f) In homing, the operator moves in the direction of the transmitter, as discerned from the receiver. As the operator comes closer to the transmitter, the signals increase. Continuing, the operator can reach the location of the transmitter.

Figure 5.4: Basics of radio-telemetry.

neither unripe, nor over-ripened, choosing between different trees and fruits, all the while being alert of competition and predators. The carnivorous feeding behaviours are much more intricate since they may require capturing of the prey through stealth, deception, stalking or chasing, and killing the prey while being careful of the prey's defence mechanisms and competition.

The anti-predator behaviours [Fig. 5.5g–i] involve being extremely mindful of the surroundings (as demonstrated by erect ears and continuous scanning of the environment), living in groups (so that the predator is confused while chasing a particular animal), running away on spotting a predator or fighting it (using instruments such as horns, antlers or hooves).

Parental behaviours [Fig. 5.5j–m] involve protecting the eggs and the young ones, bringing them food and keeping them warm and sheltered. These are very commonly observed with k − selected species.

Aggressive behaviours [Fig. 5.5n–s] are often observed when animals are defending their territory, their young ones, are competing against other animals, or are otherwise startled — and sense an upcoming harm. These involve overt displays of teeth and claws, making sounds such as roar, trumpet or growl, running towards the opponent and physically harming the opponent through teeth, claws, horns, antlers, hooves, etc.

Social behaviours [Fig. 5.5t–v] are observed in social animals such as penguins (living in colonies), lions (living in prides) and monkeys (living in troops). The social groups often have a hierarchy and a place for different animals. Those up in the hierarchy may display aggressive competitive behaviours, while those at the bottom may show submissive responses such as lying low and licking. The social groups also reveal several cooperative behaviours — as when the group is attacked and puts up a common defence, or when the young ones are taken care of by the extended family.

Mating behaviours [Fig. 5.5w,x] involve finding the mate (through scent, sound or other signals), approaching the mate, putting up a display of suitability (such as a dance or a song), fighting with competitors and copulation.

The scientific study of animal behaviour is known as Ethology. The learnings from Ethology provide important information to aid wildlife management.

5.6.1 Animal behaviour as a tool for wildlife management

The study of animal behaviour is important, even integral to the management of wildlife. The following cases represent some instances where the insights gained from studying animal behaviours are being used for guiding wildlife management:

1. The conservation of endangered species requires information about their natural behaviour — to develop effective reserves, protection measures and strategies. For instance, if a species is a migratory species, it will need to be afforded protection in more than one location. For species with large home range sizes, the size of the reserves has to be large. For species with prominent mating displays requiring an open area, we need to be mindful not to convert the reserve into a woodland. For species with basking behaviour, we need to provide a heating source in their enclosures [Fig. 10.3h].

2. Basic knowledge of reproductive behaviour helps improve captive breeding. E.g. knowledge of free choice of pair mate in biparental species such as Californian condors helps enhance their reproductive success. When we keep Californian condors in pairs made artificially, they often don't mate. But when they are permitted to form their own pairs and those pairs are kept together, there is a much greater chance of mating.

3. Species showing siblicide or infanticide are better managed by separating infants for protection. E.g. lesser spotted eagles have siblicide behaviours where the stronger sibling in a

(a) Basking behaviour of crocodiles observed in Sariska Tiger Reserve.

(b) Basking behaviour of turtles observed in Kaziranga National Park.

(c) Foraging behaviour of common ostrich (*Struthio camelus*) observed in Cape Town.

(d) Foraging behaviour of Indian grey hornbill (*Ocyceros birostris*) observed in Forest Research Institute, Dehradun.

(e) Foraging (stalking) behaviour of Indian leopard (*Panthera pardus*) observed in Sajjangarh Biological Park, Udaipur.

(f) Foraging behaviour of oriental darter (*Anhinga melanogaster*) observed in Kaziranga National Park.

(g) Anti-predator behaviour of impala (*Aepyceros melampus*) observed in Kruger National Park. A group of cheetahs (*Acinonyx jubatus*) can be observed in the background, near the vehicle.

(h) Anti-predator behaviour of zebra (*Equus quagga*) observed in Johannesburg Zoo.

(i) Anti-predator behaviour of wildebeest (*Connochaetes taurinus*) observed in Kruger National Park.

(j) Parental behaviour observed in African penguin (*Spheniscus demersus*) at 'Boulders' Table Mountain National Park.

(k) Parental behaviour observed in African penguin (*Spheniscus demersus*) at 'Boulders' Table Mountain National Park.

(l) Parental behaviour observed in Rose-ringed parakeet (*Psittacula krameri*) in Forest Research Institute, Dehradun.

(m) Parental behaviour observed in Chacma baboon (*Papio ursinus*) in Table Mountain National Park.

(n) Aggressive behaviour observed in Asiatic lioness (*Panthera leo*) in Gir National Park.

(o) Aggressive behaviour observed in African bush elephant (*Loxodonta africana*) in Kruger National Park.

(p) Aggressive behaviour observed in blackbuck (*Antilope cervicapra*) in Blackbuck National Park, Velavadar.

(q) Aggressive behaviour observed in Spotted deer (*Axis axis*) in Mudumalai Tiger Reserve.

(r) Aggressive behaviour observed in Indian rhinoceros (*Rhinoceros unicornis*) in Kaziranga National Park.

(s) Aggressive behaviour observed in African penguin (*Spheniscus demersus*) at 'Boulders' Table Mountain National Park.

(t) Social behaviour observed in African penguin (*Spheniscus demersus*) at 'Boulders' Table Mountain National Park.

(u) Social behaviour observed in Asiatic lioness (*Panthera leo*) in Gir National Park.

(v) Social behaviour observed in Rhesus macaque (*Macaca mulatta*) in Agra Bear Rescue Facility.

(w) Mating behaviour observed in Asiatic lion (*Panthera leo*) in Gir National Park.

(x) Mating behaviour observed in toads at Wildlife Institute of India, Dehradun.

Figure 5.5: Examples of animal behaviour.

clutch kills the weaker sibling. In such cases, we may help the population grow faster by artificially removing the weaker sibling and raising it away from the stronger sibling. Similarly in the case of big cats such as tigers, an incoming male may try to kill the infants sired by the resident male. Hence when we relocate tigers, we need to be mindful of the power dynamics in the area so that the young ones have a chance to reach maturity.

4. There are many social species such as golden lion tamarins where reproductive suppression of subdominant group members happens. In such cases, the subdominant members of the group are prevented from reproduction even when they are otherwise healthy. This suppression may take the form of delayed sexual maturity (puberty), inhibition of sexual receptivity, mate guarding, infanticide, etc. For these species, a strategy to increase the rate of population growth is to separate the subdominant group members. When on their own, their reproductive suppression is overcome and they become available for early breeding.

5. Nest predation and brood parasitism are being tackled by deploying territorial species for protection. In such cases, those species that are troubled by nest predation (predators eating up young ones in the nest stage) or brood parasitism (where a parasite species such as the cuckoo lays eggs in the host species' nest and the young cuckoo that emerges kills the young birds of the host species in the clutch) are given protection by nesting a species that shows strong territorial behaviour, such as the pearly eyed thrushes, near the target nest. When the territorial birds defend their areas, the nests of the target species are also automatically protected.

6. Ethology tells us that deficiency syndromes in captive reared species (like golden lion tamarins) need to be abated prior to release for successful establishment. The animals, before they are released, must *know* how to tackle the wild conditions, away from the protection of the *ex-situ* conservation areas. Often this is done through soft releases, where animals are first released into an enclosure and given time to learn the wild behaviours, before being released completely into the wild areas.

7. Many species show *imprinting* — a phase-sensitive learning in the early stages of life that help them to differentiate between own species and other species. For example, geese — which are *nidifugous* birds leaving nests shortly after hatching — need to know who their parent is, and thus should be followed for safety. To this end, they have evolved imprinting — shortly after hatching, the young goose will imprint itself to the nearest animal (or person) and identify it as the *parent*. This largely serves the purpose in the wild conditions since the parents are closest to the emerging chicks. However, when animals are captive bred (or hatched in hatcheries), this behaviour of imprinting to the nearest object may become an issue when the young bird begins to identify a human as its parent. In this case, the young bird will follow the human, and look to it for food and safety. If an animal is *sexually imprinted*, it may even identify humans as potential mates, and refuse to mate with members of its own species. It is through the study of such behaviours that we now know that imprinting needs to be avoided in captive reared species prior to release — indicating the role of deploying puppets or foster species during early development.

8. Ethology has indicated that social animals feel less stress when they are together. In many herbivores, an excess of stress may lead to a condition called *capture myopathy* [See Section 7.6], and may even result in the death of the animal. Thus when they need to be translocated, such animals should be translocated as a group for enhanced survival. This is a learning from studying animal behaviour.

9. The behaviour of animals often provides the first clues — early warnings — about environmental degradation. E.g. disruptions in swimming behaviour of minnows indicates pesticide

pollution in the water body. It is only when we make a continuous effort to monitor animal behaviours that we will get this early warning, which then helps the manager to take steps to rectify the damage, and protect and restore the habitat.

10. Behavioural indicators are used to evaluate the effectiveness of management programs and interventions. This is because population or ecosystem-level responses take long time to become evident. On the other hand, behavioural indicators — which often manifest much quickly — are extremely good proxies for the population and ecosystem-level responses that will come in the future. Thus, they are often used as potent substitutes for the evaluation of management effectiveness.

11. Understanding animal behaviours is helping to limit the impacts of humans during critical times in animals' life cycles. For instance, many species demonstrate lekking behaviour, in which they congregate during the mating season for displays. This makes them easy targets for poachers. When we have studied the animal behaviour and know that a particular species shows lekking behaviour during a particular season, we may provide it with enhanced protection from poachers during that season.

12. When we conserve wildlife, we try to conserve as much of the diversity as possible. Behaviours and cultural variations, such as specific ways of using tools by a group of chimpanzees, are also a part of this diversity. They need to be conserved as well. An animal in a zoo is not the same as an animal in the wild, primarily because it does not have the *wild* behaviours and tendencies. With this knowledge and understanding, we can make better conservation plans and strategies that also conserve *wild* behaviours and cultural variations.

5.6.2 Understanding behaviour: The cost-benefit approach

Why do lions live in groups? Why do birds build nests and devote so much time and effort to protect and feed their young ones? Why do honeybees give up their lives to protect the hive? In other words, what governs the variety of animal behaviours?

We know that the process of evolution selects those characteristics that provide an advantage or increase the 'fitness' of the species. Thus, if animals have evolved to behave in certain manners, then there must be certain advantages to the animals to behave in those manners. What are those advantages? This study of the evolutionary basis for animal behaviour — due to ecological pressures — comes under the ambit of *Behavioural Ecology*.

The fundamental tenet in Behavioural Ecology is that animals behave in a certain manner because the *benefits to the animal due to that behaviour are greater than the costs of behaving in such manner*. To understand a behaviour, we may perform a *cost-benefit analysis* — an assessment to determine whether the cost of an activity is less than the benefit that can be expected from that activity. We elucidate with some examples.

Many herbivores live in groups [Fig. 5.6a]. Let us consider the costs and benefits of living in groups. These are summarised in table 5.3. When animals live together, there are several eyes and ears on the lookout for predators. When any animal senses a predator nearby, it can alert the whole group, and the group as a whole receives an enhanced level of protection. Similarly, for animals such as zebras, living in groups can confuse the predator since all animals look similar [Fig. 5.6b]. Hence the predator will find it difficult to chase a particular animal. In cases of two species living together, such as the langurs (*Semnopithecus entellus*) and the chitals (*Axis axis*) [Fig. 5.6c], the chitals get the benefit of access to tree leaves and fruits that fall due to langur activity, and also an added protection since the langurs — at vantage positions high up in the trees — are better able to watch out for predators. At the same time, the langurs benefit from the chitals' strong sense of smell and hearing to warn against potential dangers. Thus, living in groups can reduce the predation of herbivores. However, this benefit of protection also comes at the cost of attracting predators — when there are so many prey together, the predators can easily spot the weaker members and attack them! However, this again can act to the group's advantage since over time the weaker members will be eliminated through predation, and the group will grow stronger.

Another benefit to group-living animals is an increased efficiency of foraging. This occurs because the protection afforded by the group permits each individual to devote more time to foraging. In the absence of the group's protection, each individual would have had to spend a majority of time looking out for potential dangers, leaving less time to get nutrition. However, this benefit — of enhanced foraging efficiency — also comes at a cost. Since the animals are together, there is a greater competition for food. At the same time, there is an increased risk of spread of diseases and parasites. And many parasites, especially those living in the alimentary canal, can reduce the efficiency of food absorption and assimilation.

Living together also increases access to other members of the species. Thus less time and effort need to be spent looking for suitable mates. But this comes at the cost of loss of paternity and brood parasitism — the members can cheat, and no one is the wiser about who sired whom. But this loss of individual reproduction is somewhat compensated by the help that is made available from the kin in the raising up of young ones. Since paternity is often doubtful, the extended family only knows that the kids are related to all of them — and so everyone lends a helping hand in the protection of all the young ones.

For these animals to live together, the benefits must exceed the costs.

Table 5.3: Potential benefits and costs for group living herbivores.

Potential benefits	Potential costs
	Attraction of predators
Reduced predation	1. Competition for food
Increased foraging efficiency	2. Increased risk of diseases or parasites
Increased access to mates	1. Loss of paternity
	2. Brood parasitism
Help from kin	Loss of individual reproduction

A similar cost-benefit analysis can be done for group-living carnivores such as lions [Table 5.4]. For males, the costs in terms of loss of paternity are more than overcome by the benefits of increased access to mates and protection of offsprings against infanticide. This is important because when an outsider male (or band of several males) topples a home group, it kills the cubs of the home group — to make the females receptive to new mating. When lions live together, they are in a much better position to defend the home territory, and the cubs of all of them stay protected.

For the females, the costs in terms of lower availability of food — because the food gets shared — is compensated by the benefits of joint territorial defence (which protects their cubs), and the help available from the kin in raising the cubs and in hunting for food — many lions hunting together increase the probability of making a successful kill and having food for all members of the pride.

Predators such as lions live in groups because the benefits of group living exceed the costs of group living.

The benefits of joint defence and help given in raising relatives can be explained by the concept of *kin selection*. Evolution favours those traits that increase the survival, and ultimately the reproductive success, of one's relatives. This is because of Hamilton's rule — genes increase in frequency when

(a) Studying behaviour helps explain nature. Herbivores prefer living in groups because the benefits of living together exceed the costs of living together. Here we observe a group of giraffes, zebras and impalas together in Kruger National Park.

(b) For animals such as zebras, group living is a good way to confuse predators. Here we observe a group of zebras in Kruger National Park.

(c) The langur-chital association, as seen in Sariska Tiger Reserve, is an example of two species benefitting each other.

(d) Carnivores live in groups because the benefits of living together exceed the costs of living together. Here we observe a group of cheetahs hunting together in Kruger National Park.

Figure 5.6: The cost-benefit approach to understanding animal behaviour.

Table 5.4: Potential benefits and costs for group living carnivores.

Sex	Costs of grouping	Benefits of grouping
Male	Sharing of paternity	1. Increased access to mates 2. Protection of offspring against infanticide
Female	Lower rate of food intake	1. Help from kin 2. Territorial defence

$$rB > C$$

where

r = the genetic relatedness of the recipient to the actor, often defined as the probability that a gene picked randomly from each at the same locus is identical by descent,

B = the additional reproductive benefit gained by the recipient of the altruistic act, and

C = the reproductive cost to the individual performing the act

Hamilton's rule can be rephrased as Haldane's statement: "If an individual loses its life to save two siblings, four nephews, or eight cousins, it is a 'fair deal' in evolutionary terms, as siblings are on average 50% identical by descent, nephews 25%, and cousins 12.5%." Thus evolution will select those traits that lead individuals to perform self-sacrifice to save many relatives (thus increasing the frequency of these traits in the gene pool), and through the long process of evolution, we now have the traits of altruistic behaviours for the group hardwired into the organisms.

Similarly, we can use the cost-benefit approach to explain the behaviour of territoriality in animals — a type of intraspecific or interspecific competition that results from the behavioural exclusion of others from a specific space that is defended as a territory. It aims at excluding conspecifics (or, occasionally, animals of other species) from certain areas through the use of auditory, visual or olfactory signals as well as aggressive (or ritualised) behaviours [Fig. 5.5n–s]. The cost-benefit analysis of territoriality is presented in table 5.5.

Table 5.5: Costs and benefits of territorial behaviour.

Costs	Benefits
1. Increased energy usage during display of territorial behaviour and fights	1. Winners get an exclusive access to resources
2. Increased time demands during display of territorial behaviour and fights	2. Once established, territory reduces competition, and thus, demands for time and energy
3. Increased risk of predation when individuals are busy displaying territoriality	3. Regulates the size of population, ensuring that over-exploitation of resources does not happen

5.6.3 Studying behaviour

A study of animal behaviour is often done in two ways:

1. through the use of instrumental recordings, and

2. through the use of on-site observations.

Instrumental recordings include trackers such as collars (see Section 5.5) and automatic cameras. Collars attached to the animals, especially those that also measure and record the GPS location, movement direction, speed, body temperature and heart rate, provide a wealth of information that can be used to answer questions such as:

1. When does the animal show greatest activity?

2. When does the animal sleep?

3. Which locations does an animal visit?

4. Are two animals found in the same place at the same time? (This requires collaring of multiple animals.)

However, installation of the collar requires immobilisation, capture and handling of the animals — which is difficult, and the collars have limitations — such as low battery life and influences on the animal behaviour. Thus, while they are used for studying animal behaviours, we also make use of other devices such as camera traps. These are automatic cameras with sensors (such as trip wires, passive infrared sensors, radar, sonar, etc.). Whenever an animal comes in front of the camera trap, a picture is taken or a video gets recorded. Camera traps often also record the time and the location of the picture/video. They can be left in forest areas for extended periods of time and provide a large amount of data that can be used to answer questions such as:

1. What are the times when a particular species is active? This information can be discerned from the time stamps on the photos/videos taken by the camera traps.

2. What are the locations where a particular species is active? This information can be discerned from the location stamps on the photos/videos taken by the camera traps.

3. Which species are found in a particular location? This information can be discerned by making a list of all the species recorded in the photos/videos taken by the camera traps.

4. Which species show similar activity patterns?

5. Which species show similar localisation?

6. How do animals move? This information can be discerned from the photos/videos of animals taken by the camera traps.

7. How many individuals of a species are there in an area? The identification of individual animals can be done using body patterns (such as tiger stripes or leopard rosette spots) and specific features (such as a birth mark or fight marks on the body of an individual).

Examples of activity patterns discerned from camera traps are shown in figures 5.7a–c. To make these activity pattern charts, camera traps were deployed throughout Sariska Tiger Reserve. To make the activity pattern of a species such as the tiger, all the pictures of individuals of that species taken by the camera traps are selected. Since each picture has a time stamp, we can analyse the time stamp data to get the number of pictures taken at different times, and plot them to get an activity pattern graph. Different activity pattern graphs for different species can be superimposed to study animal interactions.

Figures 5.7a and b show that predators are normally active during the night time and during crepuscular hours since a cloak of darkness helps the predators to attack their prey using stealth. Figures 5.7a and c show that preys are more active during the daylight hours. During the night time,

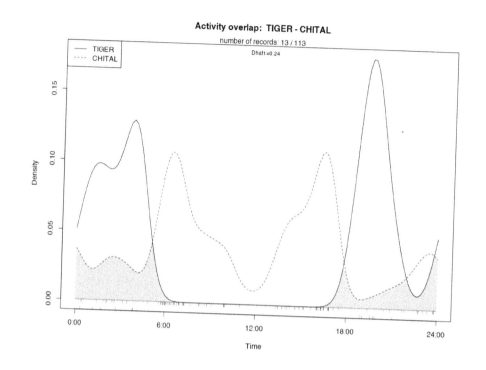

(a) Activity pattern of tiger and chital as observed in Sariska Tiger Reserve.

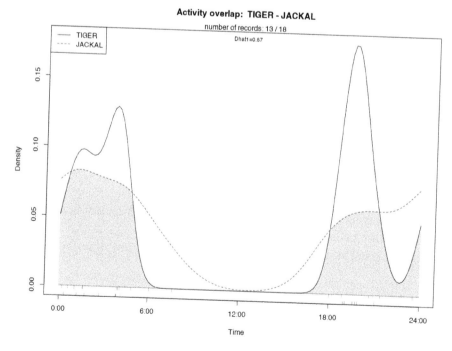

(b) Activity pattern of tiger and jackal as observed in Sariska Tiger Reserve.

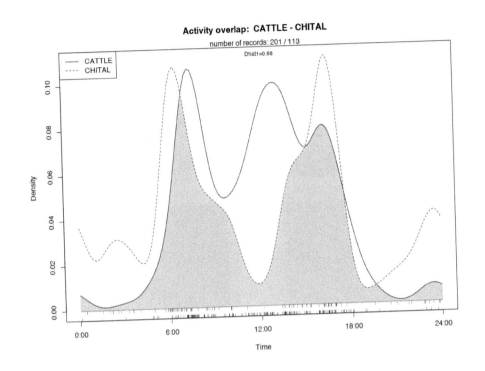

(c) Activity pattern of cattle and chital as observed in Sariska Tiger Reserve.

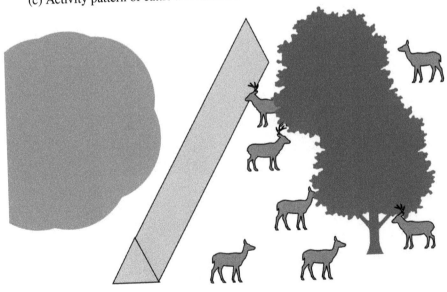

(d) Making of ethograms begins with a description of site and setting, followed by definition of various behaviours. A depiction of the site and setting of making ethograms at Sariska Tiger Reserve.

(e) Sitting

(f) Standing

(g) Walking

(h) Looking

(i) Feeding

(j) Running

(k) Auto-grooming

(l) Allo-grooming

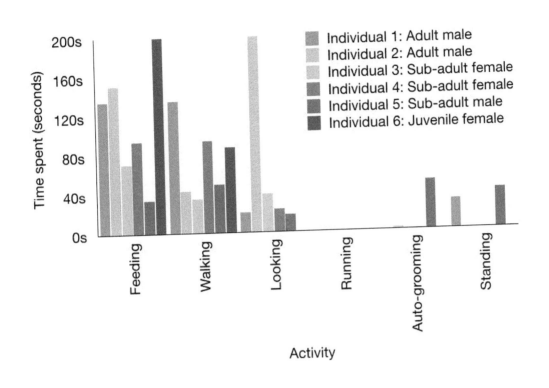

(m) Time budget graph in seconds for the study done at Sariska Tiger Reserve.

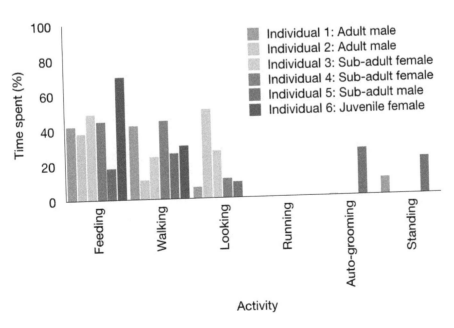

(n) Time budget graph in percentage for the study done at Sariska Tiger Reserve.

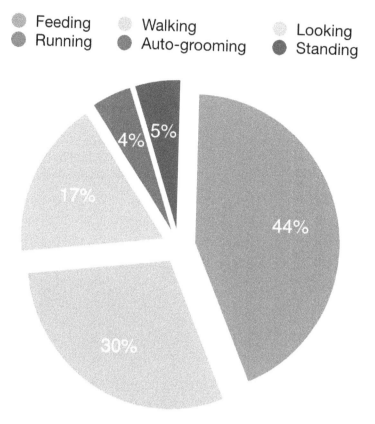

(o) Time budget pie chart for the study done at Sariska Tiger Reserve.

Figure 5.7: Some methods of studying animal behaviour.

they are better off hiding or remaining in large groups for their protection. Figure 5.7a shows that the activity patterns of predator and prey species differ from each other. During dawn and dusk, the activity patterns coalesce, providing good chances of the predators making a kill. Figure 5.7b shows that while the activity patterns of predators is similar, they are slightly different, providing for niche differentiation and a way to reduce inter-specific competition. Thus, jackals maintain a substantial activity during the day time when tigers take rest. Figure 5.7c shows that for different prey species such as the chital and cattle, the activity patterns overlap, thus providing cross-species protection from large numbers, a phenomenon we also discussed in Section 5.6.2.

On-site observations can be used to make *ethograms* — inventory of behaviours exhibited by an animal during a behaviour exercise. Making ethograms begins with a description of the study site and setting [Fig. 5.7d]. The example ethogram was made in Sariska Tiger Reserve at the site of a bund separating a water body and a woodland. Thus, this ethogram depicts the behaviour of animals — here a herd of chitals (*Axis axis*) — that have come near the water body — either to drink water, or to take a dip, or to graze in the nearby locations, or to rest in the cooler areas. Since many animals make use of the water body, this is a site where animals can encounter predators. They may also show competition behaviour when many animals are using the same resource.

Following the description of the site and setting, we need to define the behaviours. Definition of behaviours is important to pin-point the behaviour during the exercise. For instance, how do we differentiate between walking and running? Do we take a measure of speed threshold to say that any animal moving above a particular speed is running? But an accurate measurement of speed may be difficult in a field setting. At the same time, the thresholds may be different for different observers.

Thus we need to precisely define each behaviour, so that any observer can repeat the measurement. Precisely defining various behaviours also makes the observations free from observer biases. While defining, we often use such characteristics that can be readily observed, and do not need to be measured using instruments. For the example ethogram, we define the following behaviours:

1. Sitting: Abdomen touching ground, legs folded, stationary [Fig. 5.7e]

2. Standing: All hooves touching ground, legs straight, animal stationary, a sub-dominant interval during walking or feeding [Fig. 5.7f]

3. Walking: Animal moving at slow pace with at least one hoof touching ground at all times [Fig. 5.7g]

4. Looking: Animal stationary, ears raised in alert position, actively looking around [Fig. 5.7h]

5. Feeding: Mouth towards ground, eating grass [Fig. 5.7i]

6. Running: Animal moving at fast pace with at least some times where all hooves are above ground [Fig. 5.7j]

7. Auto-grooming: Scratching or licking some part of own body [Fig. 5.7k]

8. Allo-grooming: Scratching or licking some part of other's body [Fig. 5.7l]

Next, we may do a scan sampling [Table 5.6] or a focal animal study [Table 5.7].

Table 5.6: An example of scan sampling.

Obs: Ankur Awadhiya			Date: 06/12/2017				Weather: Cloudy		
TR: Sariska			Range: Sariska				Beat: Karnikavas		
Location: Water hole			Start: 15:14 hours				End: 15:47 hours		
Species: Chital			Terrain: Flat						
Time in	Adult male	Adult male	Adult male	Sub-adult female	Juvenile female	Juvenile female	Juvenile female	Time out
14:55	Walking	Feeding	Feeding	Feeding	Feeding	Looking	Walking	14:56
14:57	Feeding	Feeding	Walking	Feeding	Walking	Feeding	Feeding	14:57
14:58	Feeding	Looking	Running	Running	Feeding	Looking	Alert	14:59
15:00	Feeding	Feeding	Running	Feeding	Feeding	Walking	Feeding	15:00
15:00	Walking	Looking	Walking	Walking	Walking	Feeding	Feeding	15:01
15:02	-	Standing	Standing	Feeding	Looking	Feeding	Running	15:03
15:04	-	Walking	Feeding	Walking	Walking	Feeding	Feeding	15:05
15:06	-	Walking	Walking	Standing	Walking	Feeding	Feeding	15:06
15:07	-	Looking	Running	Feeding	Walking	Feeding	Feeding	15:08

In the scan sampling method, we note the beginning time of observation (called 'time in'), and then scan the complete group, noting the activities that different individuals are displaying at that time. Once a scan is complete, we note the time of ending the scan (called 'time out'). Then the process is repeated, to get a record as shown in table 5.6.

In the focal animal study, one individual is observed carefully for a set time (say, of 5 minutes). In this period, the times spent by the animal in different activities are recorded, as shown in table 5.7.

Once we have the scan sampling table and/or the focal animal study, we can use the data to make a time budget table, as shown in table 5.8. It shows the amount of time spent by animals in

Table 5.7: An example of focal animal study.

Obs: Ankur Awadhiya		Date: 06/12/2017		Weather: Cloudy	
TR: Sariska		Range: Sariska		Beat: Karnikavas	
Location: Water hole		Start: 15:14 hours		End: 15:47 hours	
Species: Chital		Terrain: Flat			
Individual 1: Adult male					
S. No.	Behaviour	Start	End	Time spent	
1	Feeding	15:14:40	15:15:05	25 s	
2	Walking	15:15:05	15:15:27	22 s	
19	Walking	15:19:30	15:20:00	30 s	
20	Running	15:20:00			
Individual 2: Adult male					
S. No.	Behaviour	Start	End	Time spent	
1	Looking	15:24:43	15:25:59	1 m 16 s	
2	Walking	15:25:59	15:26:09	10 s	
19	Walking	15:31:01	15:31:20	19 s	
20	Feeding	15:31:20			
Individual 6: Juvenile female					
S. No.	Behaviour	Start	End	Time spent	
1	Feeding	15:41:59	15:43:20	1 m 21 s	
2	Walking	15:43:20	15:43:45	25 s	
19	Feeding	15:46:45	15:46:47	2 s	
20	Walking	15:46:47			

different activities. The time budget table can also be converted to time budget graphs [Fig. 5.7m,n] and time budget pie charts [Fig. 5.7o].

In the example ethogram and time budget data, we observe that the dominant behaviours of the chitals near this water body were feeding, walking and looking. This is expected since water bodies support grazing and browsing areas with fresh vegetation. To feed, the animals must move. At the same time, the presence of water and prey also attracts predators, so the chitals must be on the look out for predators. Thus, they also spend a considerable time actively looking around in alert positions.

We also observe from the data that juveniles spend less time looking than adults and sub-adults. This is possibly because of parental protection that is available to the juveniles, or because the juveniles are yet to understand the dangers of this area. Being in the group, they'll learn about suitable behaviours as they grow up.

Another observation is that sub-adult males spent considerable time in auto-grooming. This could be because they are nearing sexual maturity, and may soon need to present themselves to suitable mates. Thus, this behaviour of substantial auto-grooming emerges as the males grow up.

In this way, ethograms and time-budget analyses can help us record, understand and explain the behaviours of animals.

Table 5.8: Summary of times spent by individuals on different activities.

Activity	Individual 1: Adult male	Individual 2: Adult male	Individual 3: Sub-adult female
Feeding	135 s	151 s	71 s
Walking	135 s	43 s	35 s
Looking	20 s	3 m 21 s	39 s
Running	0	0	0
Auto-grooming	0	2 s	0
Standing	30 s	0	0
Time spent	5 m 20 s	6 m 37 s	2 m 25 s
Activity	Individual 4: Sub-adult female	Individual 5: Sub-adult male	Individual 6: Juvenile female
Feeding	94 s	34 s	3 m 21 s
Walking	94 s	49 s	87 s
Looking	23 s	17 s	0
Running	0	0	0
Auto-grooming	0	50 s	0
Standing	0	40 s	0
Time spent	3 m 31 s	3 m 10 s	4 m 48 s

CHAPTER 6

Wildlife disease management

One of the objectives of wildlife management is to keep the animal populations healthy. Healthy populations have faster rates of growth, a large capacity to resist changes in their environment and enormous resilience to recuperate from changes. To this end, it is important to understand the factors that make individuals healthy and free from diseases. Keeping the populations healthy also requires that we regularly monitor them and provide with the requisite care and treatment as and when needed. In this chapter, we shall discuss the management of wildlife diseases to keep wildlife populations in the prime of health.

6.1 HEALTH AND DISEASE

Health is a state of complete wellbeing — physical, mental and otherwise. It is manifested by the ability of an animal to acquire, convert, allocate, distribute, and utilise energy with maximum efficiency, or in other words, the ability to perform its natural bodily functions optimally. A good health is of prime importance for conservation, since it reflects itself in the population growth rate.

Disease, on the other hand, is a condition of impairment of the normal functioning of the body. The word originates from the Old French *desaise* meaning lack of ease. A diseased animal will often depict a specific set of symptoms and traits that emanate from this lack of ease. These symptoms may include lack of appetite — the animal refrains from eating or eats very little, lack of energy, irritable nature, itching and aloofness. While a lack of ease may also occur due to an external injury, we traditionally exclude injury from the definition of disease. Thus, disease may be defined as "any *condition which results in the disorder* of a structure or function in a living organism that is not due to any external injury."

One aim of wildlife management is to maximise the number of healthy animals, or to maintain the population in a healthy state — by treating diseases and creating such conditions wherein diseases are unable to proliferate.

6.1.1 Kinds of diseases

Diseases may be classified in several ways:

1. On the basis of severity & duration, diseases can be classified as

   (a) Acute diseases — diseases of short durations. Acute diseases act quickly and are often of recent onset. Examples include influenza and myocardial infarction.

   (b) Chronic diseases — diseases of long duration. Chronic diseases may be persistent or otherwise long lasting in their effects and often come with time, taking a long period to manifest themselves. Examples include cancer and several deficiency diseases caused by malnutrition.

</>

2. On the basis of transmissibility, diseases can be classified as

   (a) Communicable diseases — diseases that can be communicated, or are transmissible from one animal to another animal. Most communicable diseases are infectious diseases that are caused by disease causing agents, called pathogens. Common examples are bacterial, viral, fungal and parasitic diseases caused by pathogens such as bacteria, virus, fungi and parasites. Such diseases are also known as *infections*, defined as "the development or multiplication of an infectious agent after it has entered into the body of a host." Some infections are also *contagious*, meaning that they are easily transmissible just by having contact with an ill animal or their secretions. Examples include herpes and ringworm.

   (b) Non-communicable diseases — diseases that are not transmissible from one animal to another animal. Common examples include cardiovascular disease, diabetes and malnutrition.

3. On the basis of their geographical extent, diseases can be classified as

   (a) Enzootic diseases also known as localised diseases — animal diseases peculiar to or constantly present in a locality. For such diseases, the disease causing agents (pathogens) and the conditions for their spread are always available in their localised areas, maintaining the disease in the population without the need for external inputs. An example is the enzootic abortion of ewes — late abortion in ewes caused by *Chlamydophila abortus* introduced to a flock by a carrier sheep, often one that has been brought from outside without proper monitoring, quarantine and treatment. Once the pathogen has reached the area with the carrier sheep, it is able to sustain itself in the locality, infecting other sheep. In this way, the disease continues to manifest itself in the locality for extended periods of time. And since the pathogen requires vectors and carriers to spread it to other areas, in the absence of movement of animals, the disease will remain localised and will not spread to larger areas, thus becoming an enzootic disease.

   (b) Epizootic diseases — diseases that suddenly and temporarily affect a large number of animals over a large area. An example is *Sylvatic plague*, which when it occurs, rapidly affects the majority of animals in a large area.

   (c) Panzootic diseases — epizootic diseases become panzootic diseases when they spread across a large region of Earth (for example a continent) — or even worldwide. An example is the H5N1 Avian Influenza which spread all over the world.

   (d) Sporadic diseases — diseases occurring upon occasion or in a scattered, isolated, or seemingly random way. An example is rabies in dogs, which occurs sporadically.

   (e) Outbreaks — sudden increase in occurrence of a disease in a particular time and place. An example is an outbreak of ebola. Outbreaks often come and go away quickly.

6.1.2 Disease causation factors

Diseases are caused through a requirement of, and interaction between several factors. These disease causation factors can be classified as:

1. Predisposing factors — A predisposing factor is a condition or situation that may make an animal/person more at risk (or *susceptible* to disease), e.g. heredity, age, gender, environment and lifestyle. For example, old age is a predisposing factor that makes animals more susceptible to diseases since the immunity of older animals is often less.

2. Precipitating factors — A precipitating factor is a condition that causes or triggers the onset of a disease, e.g. exposure to a specific disease, amount or level of an infectious organism, drug or noxious agent, etc. Thus an old animal will get the disease when it is exposed to the pathogen — making exposure to the pathogen is a precipitating factor.

3. Perpetuating factors — A perpetuating factor is a condition that maintains the disabling symptoms in an individual. Examples include insufficient nutrition and unhealthy surroundings. A diseased animal will be unable to fight diseases if it does not get sufficient nutrition — and the disease will keep going in its body, making non-availability of sufficient nutrition a perpetuating factor for the disease.

4. Absence of protective factors — Protective factors are factors or traits of an individual that serve as strengths or assets that can be accessed to promote a healthy adjustment to the disease and to lessen the severity of disease symptoms. A good example is a well-functioning immune system. Animals with absence of (or reduced) suitable immune response — either due to genetics or because of factors such as insufficient nutrition — are unable to fight the disease. They thus remain in a diseased state for elongated periods of time, while also serving as agents for the spread of disease to the larger population.

In most diseased populations, we may observe the concomitant action of all four disease causation factors.

6.1.3 Characterisation of disease at the level of population

Diseases have various characteristics. Some diseases make the animals sick. Others are more serious and may lead to death of the animals. We differentiate between morbidity and mortality as follows:

1. Morbidity — It is the state of the animals being diseased or unhealthy within a population. If a large percentage of animals in a population are diseased, we say that the disease has a high morbidity.

2. Mortality — It is the term used for the number (or ratio) of animals who died within a population. If a large percentage of animals in a population die due to a disease, we say that the disease has a high mortality.

Diseases are also characterised in terms of their incidence and prevalence:

1. Incidence is a measure of the probability of new infections. It reflects the probability of occurrence of a given medical condition (disease) in a population *within a specified period of time.*

2. Prevalence is a measure of the snapshot of infections. It reflects the proportion of a particular population found to be affected by a medical condition (disease).

   It is calculated by comparing the number of animals found to have the condition with the total number of animals studied, and is usually expressed as a fraction, as a percentage, or as the number of cases per 10,000 or 100,000 animals.

In other words, incidence conveys information about the risk of contracting the disease, whereas prevalence indicates how widespread the disease already is.

6.1.4 Pathogen

Pathogens are disease causing agents. The word has Greek roots: *pathos* = disease and *genes* = born of. In the case of infectious diseases, we also refer to pathogens as infectious agents — microorganisms whose presence (or excessive presence over a threshold) is required for the occurrence of an inapparent (subclinical) infection or a clinically manifest infectious disease. Some examples of infectious agents and the diseases caused by them are presented in table 6.1.

Table 6.1: Some examples of infectious agents and diseases caused by them

Infectious agent	Example of a disease caused by the agent
Bacterium	Tuberculosis
Rickettsia	Typhus fever
Chlamydia	Trachoma
Fungus	Ringworm
Parasite	Tapeworm
Virus	Rabies
Prion	Bovine spongiform encephalopathy

Pathogens can be characterised in terms of their

1. infectivity — a measure of the relative ease with which microorganisms establish themselves in a host species. Infectivity is measured as

$$Infectivity = \frac{Number\ of\ individuals\ infected}{Number\ of\ individuals\ exposed\ to\ agent}$$

2. infectiousness (or transmissibility) — a measure of the extent to which an agent can be transmitted from one host to another. Infectiousness is measured as

R_0 (Basic reproduction number) — the number of secondary infections in a susceptible population that result from a primary infection, indicating the number of individuals that are infected, on average, by each infected individual.

3. Pathogenicity — a measure of the extent to which clinically manifest disease is produced in an infected population. Pathogenicity is measured as

$$Pathogenicity = \frac{Number\ of\ individuals\ developing\ clinical\ illness}{Number\ of\ individuals\ infected}$$

The pathogenicity of an infectious agent is determined by:

(a) its ability to invade tissues (invasiveness) e.g. *Shigella*,

(b) its ability to produce toxins (intoxication) e.g. *Clostridium botulinum*,

(c) its ability to cause damaging hypersensitivity (allergic) reactions e.g. *Mycobacterium tuberculosis*,

(d) its ability to undergo antigenic variation e.g. influenza virus, and

(e) its ability to develop antibiotic resistance e.g. bacteria with plasmid-mediated genetic antibiotic resistance.

4. Virulence — a measure of the extent to which severe disease is produced in a population with clinically manifest disease. Virulence is measured in terms of

$$Virulence = \frac{Number\ of\ individuals\ developing\ severe\ and\ fatal\ disease}{Number\ of\ individuals\ with\ disease}$$

5. Lethality — a measure of the relative ease with which an agent causes death in a susceptible population.

6. Infective dose — a measure of the number of individuals of the infective agent needed to cause an infection. For example, if ingestion of around 10 bacteria of *E. coli* or around 100 million bacteria of *Vibrio cholerae* causes disease, we'll say that the infective dose of *E. coli* is 10 bacteria, and the infective dose of *Vibrio cholerae* is 100 million bacteria.

Infective dose depends both upon the route of transmission (e.g. the bacterium *Bacillus anthracis* that causes anthrax has an infective dose of 10 spores or less through cutaneous route, and an infective dose of several thousand spores through inhalation route) and the host susceptibility (the species and health condition of the individual).

6.1.5 Disease transmission

For diseases to be transmitted, pathogens must have a way to reach and enter into the body of a host, multiply inside it, and then move out to enter into another susceptible host, thus forming a chain of transmission. Pathogens can enter into the body through several portals of entry into host, such as

1. the respiratory tract,

2. penetration of intact skin,

3. the gastrointestinal tract,

4. mucous membranes,

5. the genitourinary system,

6. placenta, etc.

The transmission of diseases to these portals of entry is categorised into

1. Horizontal & vertical transmission:

    (a) Horizontal transmission is the transmission of a disease between members of the same species that are not parent and child. e.g. transmission of rabies through the bite of a rabid animal — the rabid animal may transmit the virus to *any* animal through its bite, not necessarily its parents or children.

    (b) Vertical transmission is the transmission of a disease between parent and child. Such a transmission may occur before birth (*in utero*) or immediately after birth (through ingestion of breast milk or direct contact during or after birth). e.g. transmission of chlamydia from parent to child.

    Modes of vertical transmission include:

    i. Hereditary transmission — disease is carried through genome of either parent, e.g. sickle cell anaemia.

    ii. Congenital transmission — disease is acquired *in utero* (in the uterus) / *in ovo* (in the egg) and is present at birth, e.g. congenital heart disease.

       iii. Transplacental transmission — pathogens are transmitted from mother to foetus through placenta, e.g. Herpes.

       iv. Ascending transmission — The disease moves upwards from the urethra / vagina, often resulting in a pre-term abortion.

2. Direct & indirect transmission:

(a) Direct transmission happens through touch, when the pathogen is passed from one animal to another when their bodies touch in some way. Modes of direct transmission include:

       i. touching,

       ii. biting,

       iii. sexual intercourse,

       iv. droplet projection into eye, nose, mouth, etc., and

       v. trans-placental transmission.

(b) Indirect transmission happens without touch, when pathogens are carried to an animal in some way other than actual body-to-body contact. Modes of indirect transmission include:

       i. Vehicle-borne transmission — through contaminated inanimate materials or objects (fomites) such as cage, water, and food

       ii. Vector-borne transmission, including

          A. mechanical transmission — carriage by crawling insect, flying insect, etc. without multiplication or development of the pathogen inside the vector, and

          B. biological transmission — carriage, together with propagation (multiplication), cyclic development or a combination of these (cyclopropagative mode) inside the vector.

6.2 PRINCIPLES OF DISEASE MANAGEMENT

Many diseases can be treated or cured using medicines. However, in the wild setting, the aim often is to *manage* the disease — that is, to keep its prevalence low enough for optimal population growth. This is not only because complete eradication of diseases is often difficult, requiring abundant money and effort, but also because it is often unwarranted. After all, diseases are a part of the natural ecosystem — a form of biodiversity — and may themselves be playing a role in the functioning of the ecosystem. For example, diseases are a major way in which population sizes are controlled — when the population size increases beyond a threshold, transmission of pathogens becomes much easier, and by reduced efficiency and mortality of organisms, the population size is moved back to normal. Similarly, diseased individuals — being weak — are preferentially predated upon. This not only provides easy food to the predators, but also plays an evolutionary role by removing the weaker individuals from a population. Diseases may also have a role in regulating the mixing of populations — diseases move with moving individuals, and may decimate those populations that are not immune. At the same time, eradication of certain diseases may require application of large quantities of chemical agents, which will have their own unintended impacts on wildlife. We certainly do not wish to have a situation where "the medicine is more painful than the disease," and so a restrained approach is often best suited.

Thus, the general aim of wildlife management is to keep an eye on the prevalence of different diseases in the ecosystem. Only when the diseases have turned into outbreaks, or are threatening certain priority species, should any treatment intervention be made. In other instances, keeping the animals 'healthy' by ensuring sufficient nutrition and cleanliness should be the target. In this section, we discuss the principles of disease management applicable to wildlife situations, beginning with the epidemiology triangle.

6.2.1 Epidemiology triangle

The epidemiology triangle [Fig. 6.1] is a representation of the three prerequisites for the spread of an infectious disease:

1. a pathogen that can cause a disease,

2. a host that can be infected, and

3. an environment that brings the host and pathogen together in a way that the host gets infected.

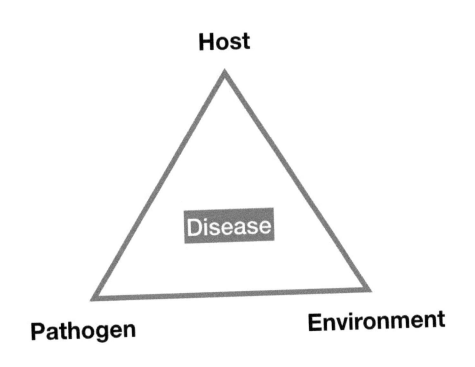

Figure 6.1: The epidemiology triangle.

In this regard, the environment plays a facilitating role. Dirty environments where the pathogens (or their vectors) can thrive, and extreme environments that weaken the immune responses of the host, create conditions where diseases can spread easily.

The epidemiology triangle hints at the steps needed to control diseases. Since diseases form through an interplay between host, pathogen and the environment, we may break the chain of infections by working on any one or more of the three: host, pathogen and the environment.

6.2.2 Host and host responses

A host is defined as an organism that harbours a guest and provides it with nourishment and shelter. In the context of diseases, the 'guest' may be parasitic, mutualistic, or commensalistic. For example, rat is a host for fleas, and fleas are hosts for plague-causing *Yersinia pestis* bacteria.

Hosts can be classified into

1. Definitive/primary hosts: A definitive or primary host is one in which a parasite reaches maturity and reproduces sexually.

2. Secondary/intermediate hosts: An intermediate host is one in which the parasite does not reach maturity or reproduce sexually.

3. Incidental/dead-end hosts: A dead-end or incidental host is an intermediate host that generally does not allow transmission to the definitive host, thereby preventing the parasite from completing its development. For example, humans and horses are dead-end hosts for the West Nile virus. People and horses can become infected, but the level of virus in their blood does not become high enough to pass on the infection to mosquitoes that bite.

4. Reservoir hosts: A reservoir host can harbour a pathogen indefinitely with no ill effects, often with important implications for disease control. This is because in the absence of ill effects, we may not even know that there are pathogens in these individuals. Thus, often the reservoir hosts get ignored during disease management strategies, and they help maintain the disease in the ecosystem.

5. Amplifier hosts: An amplifier host is one in which the level of pathogen can increase and become high enough for transmission (often through vectors such as mosquitoes). The presence of amplifying hosts helps create situations of outbreak.

Some hosts may act as carriers. A carrier is an individual that has the disease, but not its symptoms. It is capable of transmitting the disease to a new individual, e.g. a carrier of tuberculosis. Carriers may be

1. incubatory carriers — individuals who are capable of transmitting a pathogen to others during the incubation period of the disease, when the disease shows minimal or no features,

2. convalescent carriers — individuals who are fully cured of a particular disease — and thus show no symptoms — but are still capable of transmitting the disease to others, or

3. asymptomatic carriers — individuals who have the pathogen in their bodies, show no signs or symptoms of the disease, and are capable of transmitting the disease to others.

It is important to appreciate that not all individuals that are exposed to pathogens will acquire the disease. This is primarily due to host defense mechanisms which operate at different levels in different individuals. These include:

1. non-specific host defense mechanisms which are not specific to a particular disease, such as

   (a) intact skin (acting as a natural barrier),
   (b) nasal cilia (which move mucous and particles — including pathogens — towards nasal cavity and then out of the organism),
   (c) tears, saliva and mucous (which prevent drying and chapping, and contain lysozyme and other antimicrobial agents), and
   (d) gastric acid (which kills pathogens due to its low pH),

2. specific host defense mechanisms which are specific to a particular disease, such as

   (a) naturally acquired active immunity from a previous infection,
   (b) naturally (transplacentally) acquired passive immunity in newborns,

(c) artificially acquired active immunity from immunisation, and

(d) artificially acquired passive immunity from immunoglobulins and antitoxins, and

3. host responses to infection which prevent or reduce the severity of the disease, such as

(a) polymorphonuclear leucocytosis which increases the number of phagocytic white blood cells,

(b) fever which may slow the multiplication of some agents,

(c) antibody production which may neutralise some infectious agents and/or their toxins,

(d) interferon production which may block intracellular replication of viruses, and

(e) cytotoxic immune cell responses which kill cells infected with viruses.

6.2.3 Assessment of animal condition

Assessment of animal condition, especially whether an animal is diseased or susceptible to diseases due to poor condition, is necessary for wildlife health management. Often a diseased animal looks and behaves differently than a healthy animal. The diseased animal may show signs of distress — such as limping due to a sore in the leg, excessive licking or scratching of a body part that is infected and itching, difficulty in eating or swallowing, or anxiety and irritability. Some disease manifestations may be visible on the body — such as yellowing of the skin, ulcers in the mouth, dishevelled and lustreless hairs, dull skin, and lesions on the body. Behaviourally, the animal may appear aloof, inattentive, distracted, lethargic, listless, and lacking in energy. Often it will isolate itself and not be a part of the group. Due to reduced efficiency, it will, at times, show signs of malnutrition — emaciation, atrophy and protruded bones. Thus we can often tell if an animal is diseased — or in a poor bodily condition — by observing the animal and its behaviour.

While making an evaluation of animal condition, it is important to recognise that animal conditions may naturally vary over the course of life of an animal due to several causes. In particular, animal (host) condition is dependent upon:

1. the animal's age, with young and old individuals being more susceptible to diseases than adults,

2. existing diseases in the animal, both infectious and non-infectious, which may make the animal more susceptible to other diseases due to comorbidity,

3. the animal's physiological condition, such as

(a) rutting,

(b) pregnancy, and

(c) lactation

4. the environment — extreme levels of temperature and moisture may reduce host responses to infections, and

5. the environment, including the availability of food, water and cover — lack of these may increase the host's susceptibility to infections.

The assessment of an animal's condition may be done through several ways:

1. Direct assessment [Fig. 6.2] which examines several features of bad animal condition:

(a) visibility of ribs indicating poor nutrition,

(b) bony protuberance in the hind quarters indicating starvation,

(c) skin coat condition with rough, dishevelled, or hairless coat indicating underlying diseases, and

(d) bone marrow index which investigates the fat content in a dead animal's bones.

Figure 6.2: A picture of an elephant in poor body condition. Can you list the features that help recognise the poor health of this individual?

2. Population statistics, especially the quantification of natality and mortality: A high natality and low mortality indicates that the population is well fed and is doing well, whereas a low natality and high mortality may indicate that the population has prevalent diseases and/or poor availability nutrition.

3. Habitat evaluation: It investigates the availability of food, water, and shelter for animals, together with the existing levels of disturbance.

Bone marrow index

Bone marrow index is a good measure of the nutritional status/starvation of wild animals. It is derived from the sequence of energy resource utilisation that occurs when an animal faces starvation. The first resource that can be used without deleterious effects is blood glucose, and the last resource is bone marrow fat, after which the body condition rapidly degrades:

blood glucose → glycogen → sub-cutaneous fat → mesenteric fat → cardiac fat → renal fat → bone marrow fat → proteins (causing ketosis and possible death)

Since bone marrow fat is the last fat source used, it is a much better indicator of starvation than other sources. If the bone marrow fat is being used, it suggests that other sources have already been depleted due to unavailability of sufficient nutrition and/or prevalent diseases.

Bones are easily available in the field conditions — often predators feed on the flesh and leave out the bones. The species can easily be identified through their bones. Thus, an analysis of the bones can help determine the proportion of animals of various species that are facing starvation. This process does not require any intrusive mechanism to collect samples from the animals, and does not harm the animals — for they are already dead!

For the purpose of bone marrow indexing, non-haematopoietic central marrow of femur is generally utilised, for it is a large bone easily amenable to field analyses. Two kinds of bone marrow indices are used:

1. Qualitative index, based on the structure of the bone marrow: In healthy, well-fed animals, the bone marrow is solid, with a cheesy appearance. With the utilisation of the marrow fat, it becomes semi-solid, and then starts displaying a watery appearance. Thus, the bone marrow may be given the following rankings to evaluate the nutritional condition of the now-dead animal:

   - 0: degraded marrow
   - 1: watery marrow
   - 2: semi-solid marrow
   - 3: solid, cheesy marrow

2. Quantitative index, based on % fat, given as

   % fat = % dry weight of marrow − constant

   (the constant varies from 3 to 7 as per the species)

   This index requires lab analyses, and mostly suits academic purposes. It may be used to corroborate the results of the qualitative index.

The bone marrow index, especially the qualitative index, has gained popularity since it is very easy to use in field conditions. An animal is considered healthy if the index is 3 (qualitative) or % fat > 75% (quantitative). An animal is considered starved if the index is 0 (qualitative) or % fat < 25% (quantitative).

6.2.4   Interventions for the control of infectious diseases

Control of infectious diseases can be done through actions and programs directed towards

1. reducing disease incidence (new infections),

2. reducing disease prevalence (infections at a given time point) and

3. eradication of a disease.

These programs deploy three kinds of prevention:

1. Primary prevention — Control aimed at *reducing disease incidence or risk factors*, e.g. maintaining adequate nutritional status, immunisation, provisioning of clean water, and cleanliness of surroundings.

2. Secondary prevention — Control aimed at *reducing disease prevalence* by shortening the duration of the disease, e.g. prompt detection of disease, treatment with medicines, and nutritional supplementation. Secondary prevention in a group of infected animals may also result in primary prevention in the uninfected animals by reducing sources of infections and boosting immunity.

3. Tertiary prevention — Control aimed at *reducing or eliminating long-term impairments* of an infectious disease, e.g. prevention of secondary infections, management of pain, and provisioning of prosthetic implants.

The interventions aim to break the chain of infection, or the relationship between an infectious agent, its route of transmission, and a susceptible host. This can be achieved through

1. control measures applied to the host, including

    (a) active immunisation,

    (b) passive immunisation,

    (c) chemoprophylaxis,

    (d) reverse isolation — isolating susceptible animals to protect them from disease,

    (e) barriers in control situations, e.g. insect screens, air purifiers, etc., and

    (f) improving host resistance,

2. control measures applied to vectors, including

    (a) chemical treatment — insecticides, rodenticides, chemical traps, etc.,

    (b) environmental modifications — filling and draining areas with stagnant water, eliminating rodent habitats in zoos, using traps, etc., and

    (c) biological measures — introduction of predators and parasites of vectors, e.g. fish *Gambusia affinis* and fungus *Coelomomyces* to control mosquito population,

3. control measures applied to infected animals, including

    (a) chemotherapy,

    (b) isolation,

    (c) quarantine, and

    (d) restriction of activities,

4. control measures applied to secondary hosts, including

    (a) active immunisation,

    (b) isolation,

    (c) quarantine,

    (d) restriction or reduction of activity,

    (e) chemoprophylaxis, and

    (f) chemotherapy,

5. control measures applied to the environment, including

    (a) provisioning of safe water,

    (b) proper disposal of garbage, and

    (c) design of facilities and equipment, and

6. control measures applied to infectious agents (especially in controlled situations such as zoos), including

    (a) cleaning,

    (b) cooling,

    (c) pasteurisation,

    (d) disinfection, and

    (e) sterilisation.

Control of diseases also requires continuous monitoring & surveillance. Monitoring is defined as "the observation of a disease, condition or one or several medical parameters over time." It is generally done for one animal at a time, and helps ascertain the progression of disease in that animal.

Surveillance is defined as "an epidemiological practice by which the spread of disease is monitored in order to establish patterns of progression." It is done at a larger scale, and helps ascertain the progression of disease in the population.

6.3 COMMON WILDLIFE DISEASES AND THEIR MANAGEMENT

Common wildlife diseases include

1. bacterial diseases such as tuberculosis, anthrax, salmonellosis, and chlamydiosis,

2. viral diseases such as canine distemper, herpes, foot and mouth disease, rabies, and avian influenza,

3. fungal diseases such as aspergillosis, candidiasis, and ringworm, and

4. parasitic and protozoan diseases such as worms (roundworms, hookworms, tapeworms, etc.), trichomoniasis, babesiosis, coccidiosis, ticks, and mites.

We shall now explore some of these in detail.

6.3.1 Tuberculosis

Tuberculosis is a chronic and zoonotic disease caused by several species of bacteria, including *Mycobacterium bovis*, *Mycobacterium tuberculosis* and *Mycobacterium avium*.

The predisposing factors include dirty surroundings, poor ventilation, poor nutrition and interaction with domestic animals which may already have the disease.

Transmission of the disease occurs through contaminated food (including feeding on diseased animals) and water, aerosol transmission through cough and sneezes, and through direct contact.

Characteristics of sick animals include cough, difficulty in breathing, purulent discharge from nasal cavity, emaciation and weight loss, rough hair coat, alopecia (loss of hair), gastric upset, enlarged liver and spleen, enlarged regional lymph nodes that discharge pus on rupturing, and mastitis (inflammation of udders).

The diagnosis of tuberculosis is done through clinical signs, radiological examination, tuberculin skin test, smear examination of blood, and DNA amplification.

Management of tuberculosis is done via

1. surveillance,

2. quarantine: minimum 60 days or preferably 120 days,

3. long term treatment when required,

4. maintaining hygiene and sanitation,

5. destruction of (dead/severely diseased) animal and disinfection of premises,

6. treatment of priority animals with antibiotics, and

7. vaccination.

6.3.2 Anthrax

Anthrax is an acute (animals die within 72 hours of showing symptoms), highly contagious, zoonotic disease caused by the bacterium *Bacillus anthracis*. The disease is often recognised through the distinct characteristics present in the carcass of dead animals, such as

1. opisthotonus — spasm of the muscles causing backward arching of the head, neck, and spine, as in severe tetanus, some kinds of meningitis, and strychnine poisoning,

2. blood oozing from every orifice of the animal,

3. generally an absence of rigor mortis,

4. rapid post-mortem decomposition of carcass with marked bloating,

5. dark red, poorly clotted blood,

6. extensive oedema and haemorrhage in lymph nodes, and

7. splenomegaly with pulpy spleen.

Characteristics of sick animals include emaciated body, withdrawal from environment, lethargy, distinctive feeding (voraciously at certain times, and depressed and indifferent at other times), difficulty in walking (at times staggered walking), stiff-legged gait when running, swelling near umbilical regions and around genitals, and swollen (oedematous) face.

The mechanism of spread of anthrax is depicted in figure 6.3. The opening up of carcass of diseased animals (by carnivores or during post-mortem examination) leads to sporulation, permitting spread of bacteria — in the form of spores — to large areas. Thus, no post-mortem examination should be carried out for suspected anthrax cases. The spread of the spores may be through air and water, or through animals including insects, herbivores, and carnivores. Old bones often carry the spores for a long time, and complete disinfection of the area is crucial to management. Similarly, spores in the bottom sediments of water bodies may persist for long times, making regular cleaning of water holes necessary for managing the disease.

The management of anthrax is done via

1. surveillance,

2. quarantine,

3. carcass disposal through fire, *without* opening/post-mortem,

4. control over the number of scavenger animals,

5. disinfection of the site,

6. treatment of priority animals with antibiotics,

7. vaccination, and

8. construction of concrete-floored water holes that are cleaned regularly.

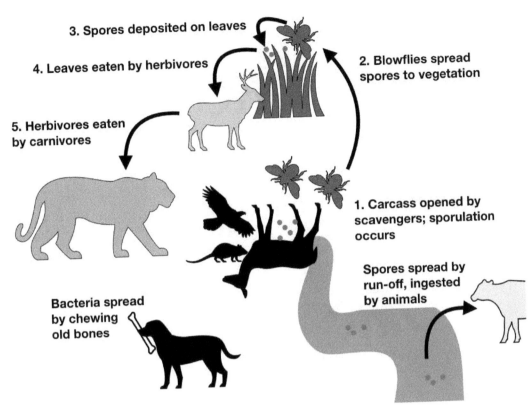

3. Spores deposited on leaves

4. Leaves eaten by herbivores

2. Blowflies spread spores to vegetation

5. Herbivores eaten by carnivores

1. Carcass opened by scavengers; sporulation occurs

Spores spread by run-off, ingested by animals

Bacteria spread by chewing old bones

Figure 6.3: The mechanism of spread of anthrax.

6.3.3 Rabies

Rabies is an acute to chronic, zoonotic disease caused by *Rabies virus (lyssavirus)*. The animal characteristics include hydrophobia (fear of water and violent muscle spasms during swallowing), drooling saliva, delirium (including avoidance, loss of fear, aggressiveness and changes in activity pattern (diurnal/nocturnal)), and unprovoked attacks.

Rabies moves through several well-documented stages in an animal:

1. Incubation stage — animal shows no symptoms

2. Prodrome stage — animal has features such as fever, anorexia, headache, pain, and numbness

3. Acute neurologic phase — animal features include anxiety, hydrophobia, delirium, hallucinations, and paralysis

4. Coma — often with symptoms including hypotension, cardiac arrhythmia, and pituitary dysfunction

5. Death

The spread of rabies occurs through the bite of an infected animal. In the wild conditions, several animals act as reservoirs of rabies [Fig. 6.4]. Since rabies may occur in a chronic form, these animals can act as rabies reservoirs for several years. When they bite an uninfected animal, the infection spreads. At times humans may also get infected through the bite of an infected wild animal, or through the mediation of the common dog which has been bitten by an infected wild animal such as a bat. Such diseases that spread from non-human animals to humans are known as *zoonotic diseases* (or *zoonoses*, singular *zoonosis*).

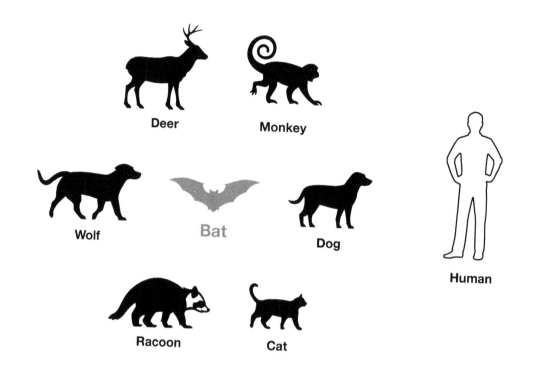

Figure 6.4: The hosts and reservoirs of rabies.

Rabies is largely incurable once the disease symptoms have manifest themselves in an animal. The management of rabies, therefore, involves surveillance, quarantine, culling of rabid animals, and vaccination, including the use of oral rabies vaccines spread in the form of food baits.

6.3.4 Ringworm

Ringworm is a chronic, contagious, zoonotic fungal disease caused by the pathogen *Trichophyton verrucosum*. The diseased animals present several disease characteristics including hairless lesions, itching, red, scaly, itchy, or raised patches that may resemble a ring, oozing or development of blisters, and thickening, discolouration, or cracking of nails.

The fungus generally spreads through direct contact with a diseased animal. The spores of the fungus may remain in the ground for a long time, also permitting soil-to-animal transmissions [Fig. 6.5].

The management of ringworm involves surveillance, quarantine, separation of diseased animals from the herd, and use of anti-fungal drugs such as ketoconazole.

6.3.5 Tapeworm

Tapeworm is a chronic, communicable, zoonotic disease caused by cestode parasites such as *Taenia solium* and *Taenia saginata*. The diseased animal exhibits loss of weight — since the parasitic tapeworm derives its food from the host animal — and presence of tapeworm fragments in stool.

The spread of tapeworm occurs through faeco-oral route. The diseased animal sheds eggs or gravid proglottids of tapeworm in its stool. Herbivores get infected upon ingesting vegetation contaminated by eggs or gravid proglottids. Once inside the body, the oncospheres hatch, penetrate the intestinal wall and circulate to muscles where they develop into cysticerci. Upon eating infected

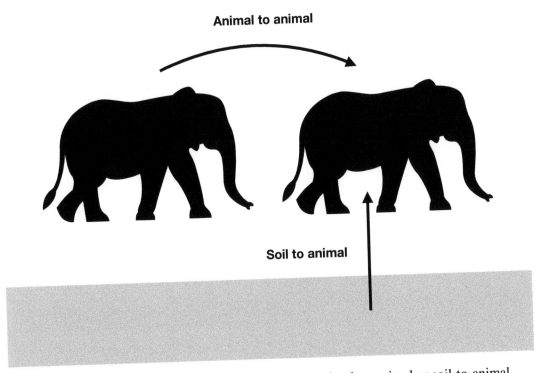

Figure 6.5: The transmission of ringworm may be animal-to-animal or soil-to-animal.

meat, the carnivores get infected. The worm gets attached to the intestine and continues to shed eggs or gravid proglottids, completing the chain of infection.

The management of tapeworm involves surveillance, quarantine, treatment of affected animals, habitat manipulation, and site disinfection/sanitisation to kill the eggs and gravid proglottids.

CHAPTER 7

Animal restraint and immobilisation

Consider the following situations:

1. An animal in the forest appears diseased and needs to be given veterinary treatment — say — a surgical operation.

2. An animal has entered human habitation and needs to be rescued and shifted back to the forest.

3. An area needs to be replenished/re-stocked with animals following a population decline. For this, the plan is to capture animals from a source site, bring them to the desired site, and release them.

4. A population is severely inbred and a genetic rescue [Ref: Section 8.4] is required, for which animals should be brought in from another population.

5. A rogue elephant is constantly attacking humans and damaging property. If it is not removed and put up for behaviour modification, there is a good chance that local villagers may kill it in revenge.

6. A research project requires tissue samples of animals for DNA profiling or disease monitoring. To take these samples, the animals need to be captured and handled.

7. An animal that needs to be tracked and monitored has to be fitted with a radio-collar for radio-telemetry-based monitoring.

8. Captive elephants are required by the department to be trained for rescue and forestry activities.

9. A ferocious animal found in a trap or a snare needs to be freed taking precautions that it does not harm humans. Thus, there is a need to tranquillise, sedate or immobilise the animal for the safety of humans during the freeing operation.

10. A 'beached' whale needs to be put back safely.

11. Orphaned leopard (*Panthera pardus*) cubs found in an agricultural field need to be removed and rehabilitated — especially since leopard is a threatened species, and in dire need of conservation.

12. A leopard has fallen into a well and needs to be taken out and rescued.

13. An endangered bird's egg needs to be removed so that a second clutch can be raised.

These are examples of some situations where capture and restraint of wild animals may be required. These operations may be done using a variety of methods — ropes and nets, devices such as traps and chutes, chemical drugs that may be injected, inhaled, or ingested, or even behavioural methods. In this chapter, we'll explore the options available, and the do's and don'ts of some of these methods.

7.1   WHAT IS RESTRAINT?

Restraint is defined as "a procedure involving capture and some degree of handling." It is of several kinds:

1. Physical restraint — only physical force is used to capture and handle the animal. Examples include hand restraint done using the handler's bare hands, with some usage of equipment including gloves, ropes, poles, shields and nets.

2. Mechanical restraint — mechanical mechanisms, such as squeeze boxes, traps, drop-floor chutes or hydraulically operated restraint chutes are deployed to capture and handle the animal.

3. Chemical restraint — drugs are used for immobilisation or tranquillisation of the animal.

4. Behavioural restraint — employing animal husbandry training, desensitisation and/or operant conditioning to perform a procedure. This is generally used in domesticated or trained animals.

In this book, we shall limit ourselves mainly to mechanical and chemical restraint. Physical restraint is only applicable to very few animals, and the care and safety precautions applicable during physical restraint are the same as those that apply during mechanical restraint. Behavioural restraint works in a very small fraction of field situations and employs topics in animal psychology which is beyond the scope of the current book.

7.2   SOME SALIENT POINTS

No matter which form of restraint is used, there are some important points to keep in mind:

1. Be mindful of the law of the land. In several countries including India, capturing a wild animal is punishable by law unless explicit prior permission has been obtained.

2. Safety is always the first consideration — safety of the personnel involved in the operation, and safety of the animal being captured or restrained. Both immediate impacts (such as death and injury to the handler or to the animal) and long-term impacts (physical, psychological and social effects on the animal) need to be considered.

3. As far as possible, try to reduce or remove unwanted stimuli. This is not only needed to calm the animal, but also because sudden stimuli may result in unpredictable responses from the animal (such as flight or fight responses, sudden jerky movement of limbs, or rapid contortions of the body), endangering the safety of the personnel and the animal. The use of devices such as blindfolds or ear plugs is, therefore, highly recommended.

4. Proceed with confidence. A lack of confidence is often perceived by animals and may increase their stress levels, or change their behaviour.

5. Plan before proceeding. All the members in the team should know why the animal is being restrained and what kinds of safety situations may arise. Issues and contingency plans need to be discussed beforehand.

6. Learn about the natural history and behaviour of the species and, if possible, of the individual animal. On feeling threatened, some animals run away, while others attack. An animal that is feeding, mating, or is with its young one may prefer to attack than to run away. An animal that has been through the process of restraint earlier may react according to its previous experience. Some macaques become too eager to be trapped if trapping is always followed by their favourite foods. On the other hand, many leopards try to avoid traps if they have been trapped before.

7. Know well the location, climate and resources. When an animal is darted, it may run away. The handler needs to anticipate this movement and ensure that the animal does not harm itself. Timely locating the animal is also crucial, not just to give antidotes and veterinary care, but also to protect the animal from being attacked by other animals. For this reason, immobilisation should not be attempted near dusk, for the animal may be lost in darkness. Knowledge about the available resources can also help tackle situations of emergency as and when they arise, for instance in cases where the animal is drowning in a water body after being darted.

8. Be prepared for emergency situations. Many things can go wrong. Someone may get hurt, the animal may get hurt, the animal may escape, the animal may die, the equipment may break down, the drugs may not work as desired due to reduced potency. Anticipate these and prepare backup plans.

9. Keep release plan on the forefront before beginning the capture process. After the animal is restrained, at some time you'll have to release the animal. It is important to do the release safely, and a release plan helps greatly.

7.3 MECHANICAL RESTRAINT

Mechanical restraint mechanisms are essentially an extension of the tools used by mankind — since antiquity — for hunting animals. These include devices such as traps and snares. The modern techniques used in wildlife management only ensure that the animal is captured without any harm (death or injury).

Just as in the case of hunting animals, for effectively deploying mechanical restraint, we need to be mindful of:

1. the nature and habits of the species — these tell us the best time for capture, the bait to be used, the trap to be used, and so on,

2. the options available for capture — the ease of availability and usage of different traps, snares, etc.,

3. the terrain on which the trap would be laid — some traps require a flat ground, while some others require an uneven surface or the presence of a tree nearby to hold the trap, and

4. the workings of the selected trap to ensure maximum effectiveness during deployment (and to troubleshoot any difficulties that may arise).

Let us explore the principles of mechanical restraint by considering a simple device — a pitfall trap. It comprises of — as the name suggests — a pit into which animal(s) fall. Thus it cannot be used for those animals that can fly, or that can jump out of the trap. We may begin by digging a pit and then waiting for animal(s) to fall in the pit. However, since "pits" are often small in comparison to the landscape — and the animals careful to avoid the pits — so the chances of animals falling into the pit are often very less. We thus need to work on increasing the efficacy of the trap. We may use methods to direct animals towards the pits — such as walls or sheets (or some kind of bait). In this case, the pitfall traps will be most effective for those animals that can be directed, cannot fly or jump high, and are small-sized (at least smaller than the size of the pits!). The animals that fit this bill are herpetofauna — small amphibians and reptiles. They are commonly found near water bodies, so that is where the pitfall traps should be laid. The presence of loose and workable soil near the water bodies will also aid in the construction and deployment of pitfall traps.

Thus, even for a simple device such as the pitfall trap, we must be mindful of the device and its limitations — the size, inability to catch flying animals, need for soft soil for digging, etc. We need to consider the location of the animals and their behaviour — such as their preference for water bodies, nocturnal activity and tendency to move along walls. And with these considerations in mind, we must make suitable modifications to the device to increase its efficiency — such as construction of walls to 'direct' animals towards the trap.

In practice, we begin with marking the layout on the ground [Fig. 7.1a]. Rainy seasons may be preferred because the animals show large movements, and are neither hibernating nor aestivating. Since the target animals prefer moving in the night, so the trap is often set up during the day time and left overnight. A direction parallel to the water body is often preferred so that animals moving both towards and away from the trap are captured. A funnel-shaped layout may also be used. Once the layout is decided, pits are dug [Fig. 7.1b]. In these pits, plastic buckets are placed [Fig. 7.1c]. The buckets should have smooth, vertical walls so that the trapped animals are unable to come out. Next, a trench is dug and stakes are put for the attachment of plastic sheets to act as directing walls [Fig. 7.1d]. The trench is so dug that the plastic sheet over the bucket crosses the bucket from middle. This ensures that animals from both sides of the plastic sheet will get trapped.

Next, the plastic sheet is attached to the stakes using ropes [Fig. 7.1e]. This is done while ensuring that the plastic sheet remains taut, thus acting like a wall [Fig. 7.1f]. Since most amphibians and reptiles prefer moving along walls, so the plastic sheet 'wall' directs the herpetofauna towards the buckets. The lower ends of the plastic sheet are tucked inside the ground [Fig. 7.1g] so that the animals are unable to duck below the plastic sheet to reach the other side of the trap. The portions of the plastic sheet above the buckets are trimmed so that the overhangs cannot be used as bridges by the animals [Fig. 7.1h].

In such an arrangement, any herpetofauna that meets the plastic sheet would — following its natural behaviour — start moving along the wall of plastic sheet. When it reaches the top of a bucket, it would fall down into the bucket and become 'trapped' since it would be unable to jump or crawl out using the steep, smooth walls. We also put leaves into the bucket to provide a hiding spot for the trapped animals and to prevent them from desiccation and direct exposure of the Sun [Fig. 7.1i]. After this, the completed pitfall trap is left to work overnight [Fig. 7.1j].

Early next morning the trap is checked for trapped animals. Since it could have trapped venomous animals such as snakes, a stick is used to prod through the leaves [Fig. 7.1k]. The trapped animals [Fig. 7.1l] are taken out, measured and documented or otherwise used [Fig. 7.1m].

A pitfall trap has several advantages:

1. It is cheap and easy to deploy. The required materials: plastic sheet, plastic buckets, rope, stakes, plough and spade are readily available.

2. It is relatively non-destructive to the habitat. The pits and trenches that are dug can easily be filled up after removing the pitfall trap.

3. It is easily repeatable for long-term studies such as changes in animal movement during

(a) Step 1: Marking the layout on the ground.

(b) Step 2: Digging of pits.

(c) Step 3: Placement of buckets to hold animals.

(d) Step 4: Digging trench and putting stakes.

(e) Step 5: Attaching a plastic sheet.

(f) Step 6: Tightening the plastic sheet.

(g) Step 7: Tucking the lower end of the plastic sheet into the ground.

(h) Step 8: Trimming off the overhangs on top of the buckets.

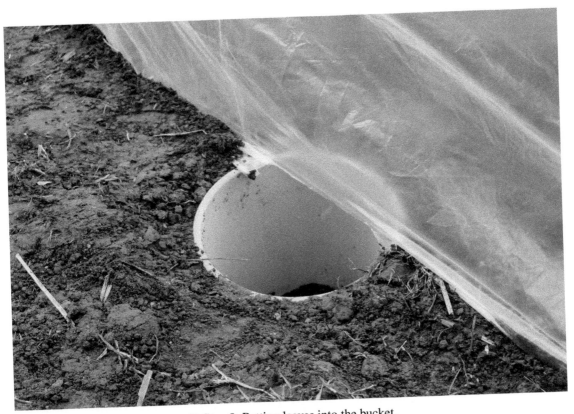

(i) Step 9: Putting leaves into the bucket.

(j) The completed pitfall trap.

(k) Step 10: Using a stick to prod through leaves looking for animals.

(l) An animal found captured in the pitfall trap.

(m) Step 11: Taking measurements and documenting the findings.

Figure 7.1: Deployment of a pitfall trap to capture animals.

different months, or over several years, or after execution of some work such as establishment of a factory.

4. Since it uses the natural behaviour of herpetofauna to follow walls, it is especially effective for these target animals.

5. Since the animals are captured live, they may be used for population supplementation or genetic rescue at other sites. They may also be tagged and used for capture-recapture studies.

However, the pitfall trap also has some disadvantages:

1. It requires lots of manual work in the form of digging pits, digging trenches, placing stakes, tying plastic sheet and tucking it in. Thus, setting up a pitfall trap needs lots of time and people.

2. The trap may require time for animals to adjust. This is especially true for those animals that avoid areas that have been recently disturbed.

3. The traps need to be checked daily to prevent animal death through heat or lack of food and water, or drowning if it rains. This becomes an issue if several traps have been deployed over a large area.

4. It is only useful for certain species, not all. It works only for those animals that cannot fly or jump high and are small in size.

5. If prey and predator fall together, the prey may get eaten, thus altering the results about animal counts.

Other devices used for mechanical restraint include metal cages [Fig. 7.2a,b] using trip mechanism to close the door [Fig. 7.2c] or a gear system to move a wall and restrain an animal [Fig. 7.2d]. Metal transport cages [Fig. 7.2e] also restrain animals during transport to a holding facility [Fig. 7.2f], or till their release. Nooses made out of ropes may be used to capture and release animals such as crocodiles [Fig. 7.2g,h]. Foothold traps and snares, though sometimes used for canids, have lost appeal with the development of better drugs for chemical immobilisation. Glue traps, deadfall traps and conibear traps often kill the restrained animals, and thus have fallen out of fashion. Nets are still used for capturing birds, monkeys and pigs, often in large numbers.

Irrespective of which mechanical device is deployed, one important consideration during usage is that animals often face distress when they are restrained. Behaviours such as biting of cage bars, hitting against the cage walls and rapid pacing are often observed. At times, the animals may injure themselves — even grievously. Thus several points must be kept in mind when using these restraints:

1. The cages should be completely enclosed to minimise outside stimuli, but with sufficient number of holes for ventilation.

2. As far as possible, iron rods should not be used since animals may break their teeth biting the rods in an attempt to escape.

3. Old and rusted cages must never be used. Any rusted portion of a cage must be replaced.

4. If a trip mechanism is used to shut the door of the cage, adequate leeway must be made to ensure that the tail of the animal does not get slammed against the door. A gap of around 1.5 inches is advisable.

5. The dimensions of the cages should be appropriate for the target animal.

6. Lightweight fibreglass cages and collapsible cages should be preferred for ease of field deployment.

7. If the animal is to be retained for long, it should not be kept in the cage, but must be moved to a separate holding facility for long-term captivity.

8. Multiple animals must not be trapped or kept together since they may injure each other in the stressful condition. This is especially important for male animals since they may fight and harm each other.

9. Adequate provisioning of food and water must be made available, including water for cooling the cage during transportation of animals.

10. Cages should not have soft materials such as rubber or padding that may be ripped and eaten by the animals.

11. For species that exhibit capture myopathy [Section 7.6], the stress levels should especially be kept minimum. This may require transporting complete herds together in dark and muffled transport containers, often with administration of suitable anxiolytic drugs.

These days chemical restraint is often preferred since the chemically immobilised animals are unable to harm themselves.

(a) Cages for capturing large mammals in Gir National Park.

(b) A cage for capturing crocodile in Kruger National Park.

(c) Inside of a capturing cage showing the trip mechanism. When the animal reaches the far end of the cage, its weight causes the trip mechanism to close the door. The platform at the far end is to tie a bait to lure the animal.

(d) A squeeze cage in Gir National Park used to handle lions.

(e) A transport cage in Gir National Park. Notice the use of large and small mesh. The cage is covered with a green sheet while in use to calm the animal.

(f) The ramp in this holding facility in Gir National Park allows smooth release of animal from the cage.

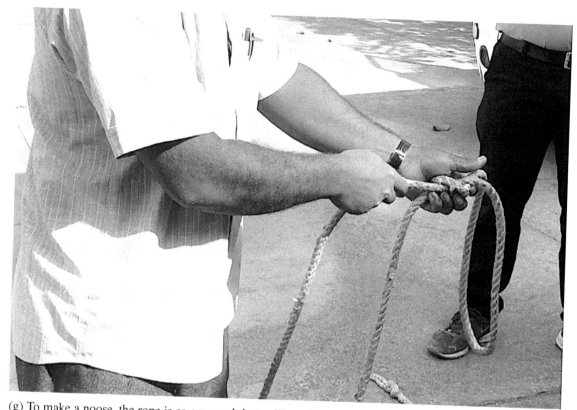

(g) To make a noose, the rope is so arranged that pulling of one end tightens the noose, while pulling of the other end opens the noose to release the animal.

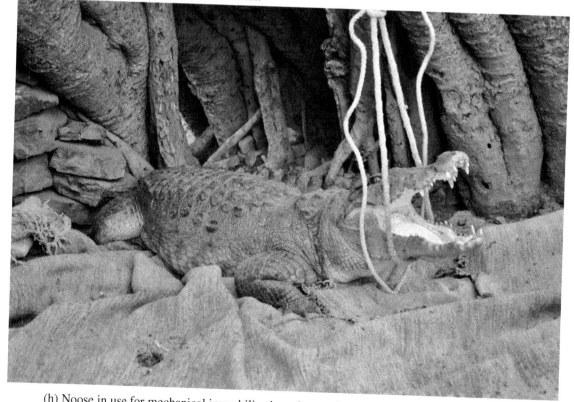

(h) Noose in use for mechanical immobilisation of crocodile in Ranthambhore Tiger Reserve.

Figure 7.2: Some examples of tools of mechanical capture and immobilisation.

7.4   CHEMICAL RESTRAINT

Chemical restraint uses chemical drugs to immobilise animals. Often these drugs are injected using darts, but oral or inhalation routes may also be used.

There are two modes of action of drugs that are employed for chemical restraint:

1. Neuro-muscular blocking: These mode uses paralysis of voluntary muscles of the body, because of which the animal is unable to move its body. Often the drugs do not have any sedative (calming and inducing sleep) or analgesic (reducing pain) properties, because of which the animal remains conscious and sensitive to pain, fear, stress and stimulation. Thus, even though the animal is unable to move its body, it can sense its surroundings and undergoes tremendous stress. The excitation of the animal can lead to release of large amounts of adrenaline and production of severe side-effects such as capture myopathy [Section 7.6]. It is for this reason that these drugs, while available as an option, are not generally preferred. When they are to be used, they are given as a combination with CNS depressants to reduce the side-effects of excitation.

2. CNS (central nervous system) depression: This mode uses production of anaesthesia (loss of sensation and awareness). Some drugs such as diazepam and xylazine reduce CNS activity, while some other drugs such as ketamine and nitrous oxide produce dissociation through hyper-excitability, leading to a trance-like state.

The common drugs in current use for chemical immobilisation can be classified into

1. opioids/narcotics (morphine and morphine derivatives) such as morphine, etorphine, and fentanil,

2. sedatives including xylazine and medetomidine,

3. dissociatives (produce analgesia and psychosis) such as ketamine — often used with sedatives to reduce side effects, and

4. tranquilisers/anxiolytics e.g. diazepam.

The modes of administration can be classified as

1. oral in which the drug is given in food or water, e.g. diazepam,

2. inhalation in which the drug is released in the air and acts upon being inhaled by the animal, e.g. chloroform, and

3. injection in which the drug is injected into the body of the animal. This is the most common mode of drug administration. The injection may be done using

   - a hand-held syringe which is often used when the animal is easily approachable or can first be handled using behavioural restraint,
   - a jab stick using which the injection can be done at a distance, and
   - darts which can be propelled with air-activated discharge mechanism, or with an explosive discharge mechanism.

Due to their universality and easy usage, darts are the preferred mode of drug delivery for wild animals. The darts may be projected using:

1. blow pipe, using air pressure from the lungs,

2. air gun, using air pressure from a pressurised canister or an air pump, or

3. gunpowder-charged gun, using gas pressure from combustion of gunpowder.

The common equipments used for chemical immobilisation are depicted in figure 7.3. The immobilisation drugs, antidotes and emergency medicines [Fig. 7.3a] are transported in an ice box [Fig. 7.3b] to keep them cool and maintain their potency.

A blow pipe is essentially a long tube with a mouth-piece. The dart is inserted at the breech-end of the tube. The mouth-piece is covered with one hand to make a good seal with the mouth and the other hand is used to take aim. Then after a deep breath, air is quickly blown into the mouth-piece [Fig. 7.3c]. The pressure of the air projects the dart towards the target.

The blow pipe has two major shortcomings — the range is less, and the projection is not accurate. These can somewhat be improved using a longer blow pipe [Fig. 7.3d]. However, for better range and accuracy, an air gun needs to be used.

The air gun consists of a barrel, a butt, a sighting mechanism, a trigger and a mechanism of using pressurised air or gas for projection. It may be a single-barrelled gun [Fig. 7.3e] or a double-barrelled gun [Fig. 7.3f]. A mechanical pump may be used in place of the pressurised gas canister [Fig. 7.3g]. A pressure gauge helps to select the correct pressure for the given animal and distance [Fig. 7.3h].

For animals with thick skins — called pachyderms — such as elephant, rhinoceros and wild buffalo, a gunpowder dart gun [Fig. 7.3i] is used in the place of an air gun for better penetration power. It consists of a barrel, a butt, a sighting mechanism, a trigger and a hammer or striker to fire the projectile. A sight adjustment is often provided for 'zeroing' the gun [Fig. 7.3j].

A typical dart [Fig. 7.3k,l] consists of a drug chamber, a needle, a piston to push the drug in the chamber towards the needle, a mechanism for pushing the piston (either pressurised air/gas or gunpowder) and a stabiliser/feather to stabilise the dart in flight. Metal darts are used in gunpowder guns and plastic darts are used in air guns and blow pipes. The needles used may be plain, collared or barbed [Fig. 7.3m].

In a gunpowder dart gun, different charges are used for the dart (called syringe charges) [Fig. 7.3n] and the rifle (called rifle charges) [Fig. 7.3o]. The rifle charge is used to push the dart towards the target animal, and the syringe charge is used to push the piston inside the dart when the dart has hit the target animal. An extractor helps position the rifle charge and aids in its removal after firing [Fig. 7.3p].

Since the drugs used for immobilisation are potent chemicals that can get absorbed through skin, gloves are essential during dart preparation and handling to avoid any mishaps. The dart for an air gun is prepared as shown in figure 7.3q–y. Plastic darts are used since they are light-weight. The materials — including the drug, the air release pin, silicone sleeve, needle, protection cover, dart, pressurisation syringe and a disposable syringe for filling the drug — are gathered [Fig. 7.3q]. The pressurisation syringe is used to ensure free movement of the piston [Fig. 7.3r]. At the rear end of the dart, a one-way valve is present to contain the pressurised air. The air release pin is used to manually operate this valve to move the piston back.

After moving the piston all the way back, the drug is loaded using the sterile disposable syringe [Fig. 7.3s], and the needle with the silicone sleeve is attached using pliers [Fig. 7.3t]. The silicone sleeve is positioned [Fig. 7.3u] and the protection cover is attached to protect against leaking and spillage during the pressurisation stage [Fig. 7.3v].

The rear end of the dart is pressurised using the pressurisation syringe [Fig. 7.3w]. The red-coloured one-way valve ensures that the pressurised air is contained behind the piston, and is unable to leak out. While the air keeps pushing the piston, the silicone sleeve covering the needle vents prevents any movement. The last step is to attach the stabiliser feather that stabilises the dart during flight upon being fired [Fig. 7.3x].

When the dart is fired and hits the animal, the needle penetrates the animal skin. The silicone sleeve is pushed back during penetration of skin, and the needle vent becomes open. This permits

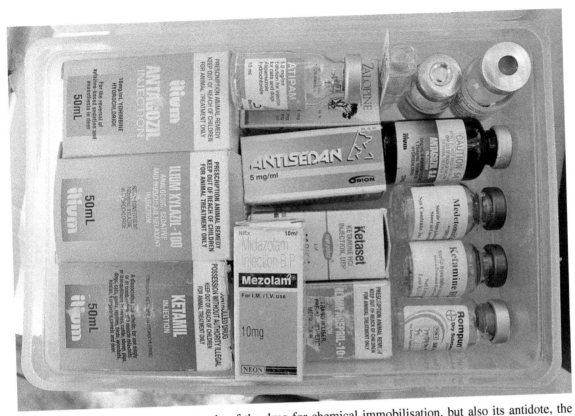

(a) The drug box must consist not only of the drug for chemical immobilisation, but also its antidote, the human antidote and drugs for dealing with emergencies.

(b) The drug box is kept cooled in an ice box. Maintenance of cold chain ensures that the drugs retain their potency.

(c) Method of holding a blow pipe. The blow pipe is a simple instrument for projecting darts using air pressure from the lungs.

(d) Longer blow pipes may be used for increased range.

(e) A single-barrel immobilising dart gun. Note the brass-coloured pressurised gas canister and the knob to increase or decrease gas pressure.

(f) Close-up of a double-barrel immobilising dart gun. Note the barrel selection knob.

(g) A mechanical pump that may be used in place of the gas canister.

(h) The pressure gauge helps to select the correct pressure for the given animal and distance.

(i) A gunpowder dart gun is used for pachyderms. The image shows the correct way of holding a dart gun.

(j) The sight adjustment helps in 'zeroing' the gun to ensure that the telescopic sight shows the point where the dart will actually go. It is often with markings for distance to the animal, that needs to be selected.

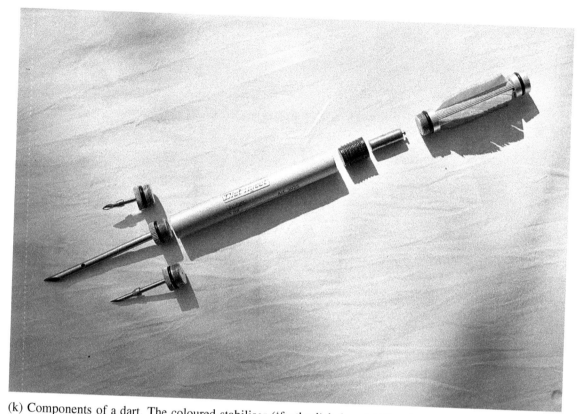

(k) Components of a dart. The coloured stabiliser ('feather') helps stabilise the dart in flight. The syringe charge (brass-coloured) explodes when the dart hits the animal. This explosion pushes the rubber piston (black) forward through the medicine chamber (silver), pushing the medicine through the needle (silver coloured).

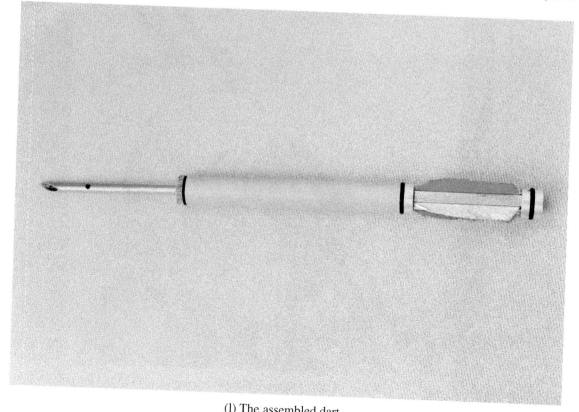

(l) The assembled dart.

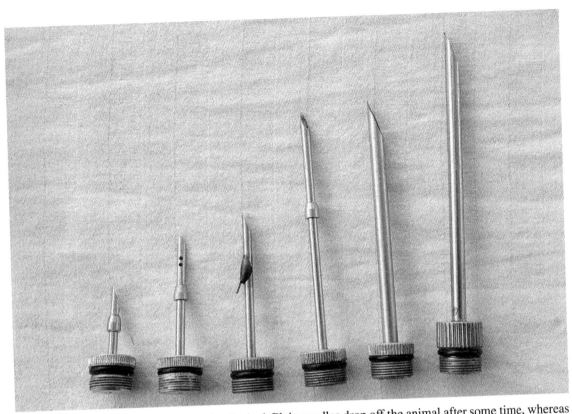

(m) The needles may be plain, collared or barbed. Plain needles drop off the animal after some time, whereas the collared (with conical thickening in the front) and barbed (with the hook in the front) needles stay on the body of the animal till they are removed. The barbed needles may require a small incision (surgery) to remove from the animal.

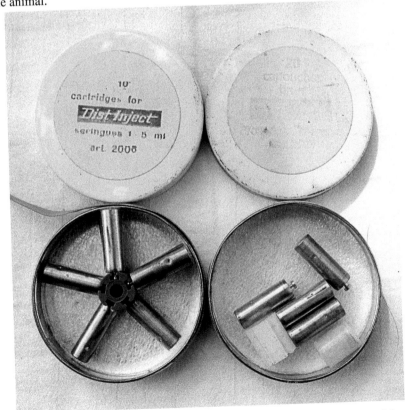

(n) Picture of syringe charges. Note the projection at the primer end. The flat shape with rod increases the pressure to ensure that the gunpowder gets ignited when the dart hits the animal.

(o) Picture of rifle charges. Note the differences from syringe charges.

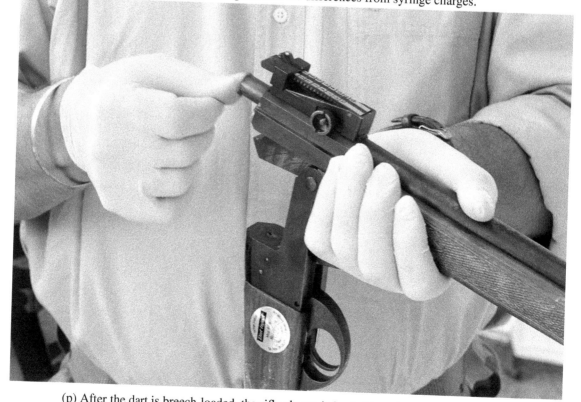

(p) After the dart is breech-loaded, the rifle charge is loaded together with the extractor.

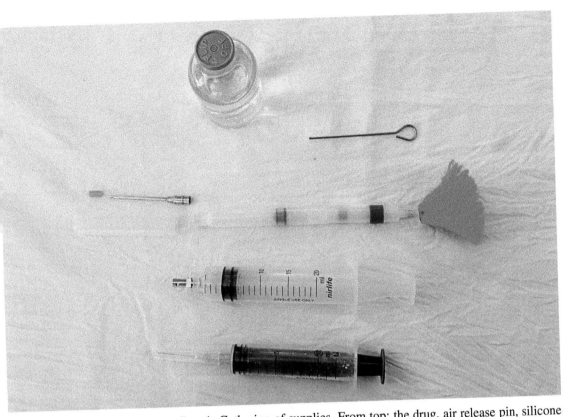

(q) The steps for preparing dart. Step 1: Gathering of supplies. From top: the drug, air release pin, silicone sleeve, needle, protection cover, dart, pressurisation syringe, disposable syringe for filling drug.

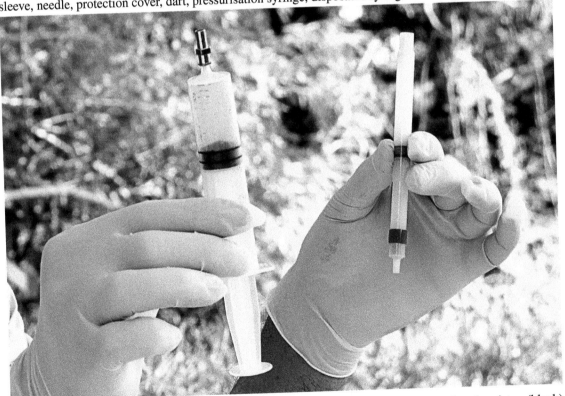

(r) Step 2: Use pressurisation syringe to push air from both sides of dart to ensure that the piston (black) moves freely. If needed, use the air release pin if the vent is blocked.

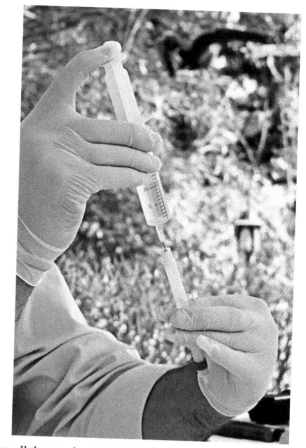

(s) Step 3: Move the piston all the way back. Holding dart in an upright position, fill the drug from the front.

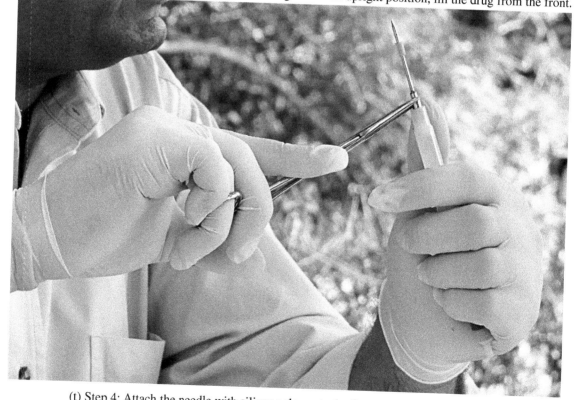

(t) Step 4: Attach the needle with silicone sleeve to the front of the dart using pliers.

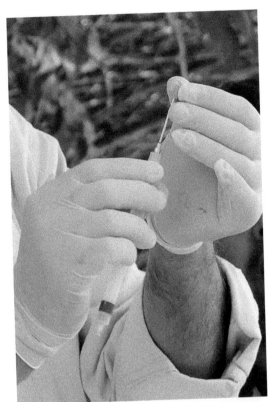

(u) Step 5: Before pressurising, double check that the silicone sleeve properly covers the vent of the needle.

(v) Step 6: Attach protection cover. The protection cover ensures that should the needle or the silicone sleeve dislodge during the next stage (pressurisation), the drug does not splash on the user.

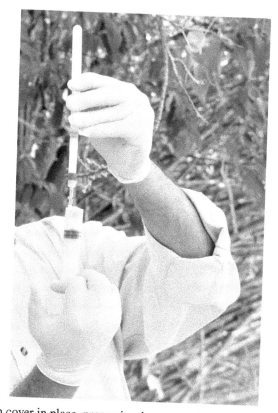

(w) Step 7: With protection cover in place, pressurise the rear end of the dart using pressurisation syringe.

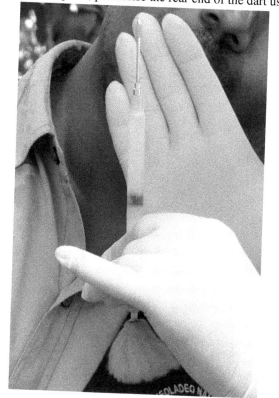

(x) Step 8: Attach the stabiliser ('feather') to the rear end of the dart. The picture shows an assembled dart.

(y) When the dart hits the animal, the silicone sleeve moves back due to impact and the piston moves forward due to pressurised air, pushing the drug into the animal through the needle vents.

Figure 7.3: Equipments for chemical immobilisation and the process of dart preparation.

the drug to come out. The pressurised air pushes the piston, emptying the drug into the animal's body [Fig. 7.3y].

The actual process of darting an animal in the field situations not only needs to take into account the scientific perspective, but also the ground conditions [Section 7.5]. Typical darting operations start early in the day when ambient temperatures are low. Evening hours are generally avoided because if the animal runs away after being darted, the cloak of darkness may hamper its search in the forest. The preparation stage [Fig. 7.4a] is the most important stage because if some supplies are left out or preparation is incomplete, coming back from the forest during the darting operation may be difficult. In the preparation stage, the supplies, people, devices, vehicles, etc. are gathered and checked. Are the equipment working properly? Are all the drugs available in sufficient quantity? Is any drug expired? Do we have multiple backups of equipment and vehicles? Has the animal been selected and located? These are some issues that are addressed in the preparation stage. Typically these are done in a camp or office where supplies are stored. At the same time, staff in the field keep an eye on the target animal and continuously provide information about its current location.

The next stage is reaching the field and briefing the team [Fig. 7.4b]. Maps, especially terrain maps are detailed and positioning of vehicles and personnel is discussed. It is important to take note of those areas where the animal may injure itself or drown. The safety precautions are re-iterated. Concomitantly documentation begins [Fig. 7.4c].

Next, the dart, antidotes and emergency drugs are readied [Fig. 7.4d] and the animal is approached with equipment. The approaching may be done on elephant back [Fig. 7.4e], on foot, or in vehicles such as helicopters [Fig. 7.4f]. Due to the sound of the vehicle, or upon sensing movement nearby, the animal may go into hiding [Fig. 7.4g] or run away, and in such cases some combing of the area may be required. When the animal is clearly visible, it is aimed at and darted

[Fig. 7.4h]. The lateral shoulder, neck, or rump of the animal is preferred for darting to deliver the drug through intramuscular (IM) route.

The animal — upon being hit by the dart, and especially when the drug begins to act — starts to feel uneasy and may start to run. The unease is exacerbated by the realisation that it is losing control over its own body. In such situations, it is best to allow the animal to settle down on its own by providing it with a calm environment without excessive noise and hustle. The use of a dart with radio transmitter is a great help because the animal does not need to be tracked continuously in real time. After around 10–15 minutes, when the drug shows its full impact, the animal sits down and loses consciousness. The caring of the animal in this stage is discussed in Section 7.5.

7.5 MANAGEMENT OF EMERGENCIES

Animal restraint and immobilisation are specialised procedures, and care is essential to prevent harm, both to the animal and to the handler. We look at safety concerns and their handling in the following subsections.

7.5.1 Animal safety

The stress faced by the animal during the process of capture, restraint, immobilisation and handling causes the activation of the sympathetic nervous system and the release of adrenaline which leads to several physiological responses in the body, including increased heart rate, bronchodilation, increased respiratory rate, glycogenolysis, lipolysis, muscle contraction, and contraction and dilation of blood vessels. When uncontrolled, an excessive release of adrenaline may lead to irreversible damage to the animal's body.

Similarly, the use of drugs for chemical immobilisation requires ample precaution, since drugs are potent chemicals and things may go wrong, sometimes very wrong while using drugs for immobilisation. Improper use may even lead to the death of the animal.

This is why all through the process of restraint and immobilisation, continuous monitoring is crucial to assess several parameters such as

1. physiological changes, including

    (a) hyperthermia, or increased body temperature,
    (b) hypothermia, or reduced body temperature,
    (c) arrhythmia, or irregular heartbeat,
    (d) stress, and
    (e) sweating,

2. irreversible injuries such as

    (a) capture myopathy [Ref: Section 7.6],
    (b) self-trauma during immobilisation and recovery when the animal is still under the effect of drugs,
    (c) fracture due to uncontrolled body movements, and falling, especially in areas with steep slopes,
    (d) drowning in vomitus (matter that has been vomited) which can lead to fatal aspiration pneumonia, and
    (e) prolonged cardiac or respiratory depression due to action of CNS depressing drugs,

3. anaesthetic depth (degree of CNS depression produced by the anaesthetic drug) which depends upon

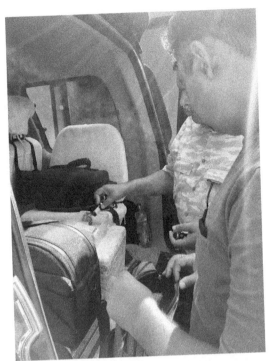

(a) Step 1: Preparation: gathering supplies, people, devices, vehicles, etc., reconnaissance, selection and location of target animal.

(b) Step 2: Reaching the spot and briefing the team.

(c) Step 3: Documentation, especially video documentation for training and legal purposes.

(d) Step 4: Preparation and readying of dart, antidotes and emergency drugs.

(e) Step 5: Approaching the animal. This can be done on elephant back, on foot,

(f) or using a vehicle such as a helicopter.

(g) Step 6: Locating the animal. The animal may be easily visible, or may be hidden.

(h) Step 7: Darting. A darted elephant in Kruger National Park.

Figure 7.4: The process of darting an animal.

    (a) the drug being used,

    (b) the dosage given to the animal,

    (c) species,

    (d) presence or absence of diseases, and

    (e) physiological status of the animal, and

4. the effectiveness of supportive care being provided to the animal.

It is especially important to continuously monitor the airways, breathing and circulation (remembered as ABC).

The monitoring methods include

1. auscultation, or listening to sounds of heart, lungs, etc. using a stethoscope,

2. monitoring the colour of mucosa, especially in the mouth, where

    (a) pink indicates healthy circulation and oxygenation,

    (b) pale colour indicates anaemia, and

    (c) blue or purple colour indicates hypoxemia — the animal is not getting sufficient oxygen,

3. capillary refill time, which gives an indication of the peripheral tissue perfusion (The method is to press mucous membrane with fingers until blanched, and note the time to return to original colour. The time to return to original colour should be less than 2 seconds, or it may indicate poor peripheral tissue perfusion.),

4. blood pressure,

5. heart rate,

6. ECG,

7. doppler flow detection of blood flow,

8. pulse oximetry to measure blood oxygenation,

9. capnography — measurement of CO_2 concentration in expired gases,

10. measurement of body temperature using thermometer, and

11. dehydration — this can be judged by gently grasping a fold of skin and rolling it between fingers; normally hydrated skin moves easily, while dehydration results in a sticky feeling. Dehydration also causes dry mucous membranes, sunken eyes, and decreased tear production, and can be identified by these features as well.

Other points of care include

1. Timing of operation: When the animal is anaesthetised/immobilised, the ambient temperatures should not be too hot or too cold. Thus mornings and evenings may be preferred, except in areas with uneven terrain where mornings are preferred to avoid missing the animal in the cloak of darkness. If necessary, the body of the animal should be cooled by spraying water, or warmed by using rags or blankets.

2. Site selection: Flat, even ground is preferred for immobilisation, so there is less chance of the animal falling and injuring/fracturing itself.

3. Control over stimulation: Eyes and ears should be kept covered to reduce stimulation from the environment. Loud noises should be avoided near the immobilised animal, for if it is able to sense the surrounding, it may get into heavy stress.

4. Breathing: Ensure that the mouth/nose is open for breathing, and see that the tongue is not blocking the airway.

5. Recumbency (act of lying down): When the drug acts, the animal lies down. There is a specific position for each species that minimises harm. If the animal has lied down in an incorrect recumbency, it should be corrected. In the case of elephants, sternal recumbency (lying down over the chest) may increase pressure on lungs and heart, which may prove fatal. Elephants, thus, must be rolled into lateral recumbency (lying down on the sides), or handled while they are in a standing position. In the case of ungulates such as deer, sternal recumbency is safer.

6. Sample collection and marking/tagging: Tissue sample may need to be collected, and the animal may need to be marked with PIT (passive integrated transponder) tags [Fig. 7.5] for later identification. These must be done in a clean and sterile manner, to avoid infections.

7. Tranquillisation: If the animal shows indications of stress, appropriate amounts of tranquillisers may need to be administered to calm the animal.

8. Inspection of equipments: The equipments, especially cranes and bulldozers must be inspected for sharp edged and appropriate functioning to ensure that they do not cause injury to the animal.

9. Transportation crates and vehicles: These must be of adequate sizes with proper provisioning of ventilation holes, inspection holes, bedding, food and water (if needed). Areas with rust and sharp edges should be promptly replaced.

10. Minimal handling: Handling of the animal must be minimised, and the transportation time should be the least required. If possible, animals may be transported in groups to maintain cohesion and reduce stress.

11. Screening of handlers for diseases: Some diseases such as tuberculosis and fungal infections can spread from humans to animals in a process known are reverse zoonosis. Thus it is important to regularly screen handlers for diseases, treat them and take steps to prevent spread of diseases to wildlife.

12. Post-release monitoring: After release, the animal must be monitored for at least 3–10 days to ensure that it has recovered well and shows no signs of the chemicals used, or of capture myopathy [Section 7.6].

7.5.2 Human safety

There are several human safety concerns during wildlife restraint, immobilisation and handling:

1. Environmental risks

   (a) Inclement or extreme weather, especially during capture and release operations. While out in the field and on elephant back, humans get exposed to the Sun, and there is a risk of getting sun-burn or heat-stroke. At times when it rains, people may get drenched and fall ill.

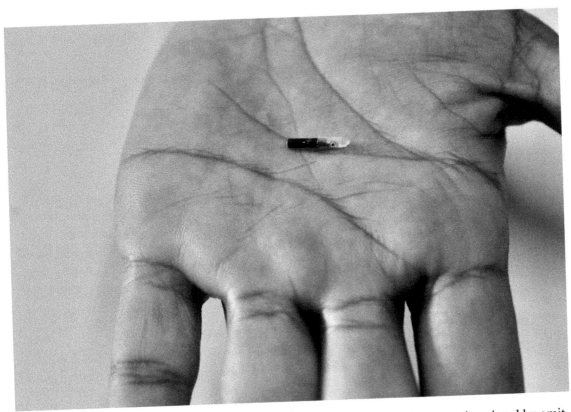

Figure 7.5: A PIT tag is an integrated transponder that responds to an interrogating signal by emitting an identifying signal, without the need of a battery.

    (b) Risks from surroundings: While on elephant back, there is a risk of getting brushed against thorny vegetation leading to scratches and injuries. While following the target animal, one may fall down and hurt oneself. There may also be a risk of drowning or getting lost in the forest.

To manage these risks, it is imperative to carry sufficient protective clothing, ropes or harness. Dangerous locations should be studied beforehand. One must also retain adequate stocks of food and water supply, together with communication systems and backup(s).

2. Disease risks

    (a) Zoonotic diseases: Zoonoses are infectious diseases that jump from non-human animals to humans. Examples include various influenzas, anthrax, brucellosis, and rabies.

    (b) Malaria, ticks, and fleas: Often forest areas are infested with mosquitoes, ticks, and fleas. Working in forest areas exposes humans to their bites.

Thus it is extremely important to use prophylactic medicines and protection such as thick clothing and insecticides.

3. Equipment related risks: Rifles and pistols are firearms, and may cause severe trauma, even death, if misused. Similarly traps, snares, and nets may not function as desired and lead to accidents. Hence there is a need for appropriate training and to follow firearms and equipment safety rules.

4. Drug-related risks: Drugs are potent chemicals and may get absorbed through the skin or through a prick. Thus several precautions are necessary to manage drug-related risks, including

(a) donning of protective gear including gloves, apron/coverall, goggles and boots,

(b) covering the dart before pressurising to prevent spills,

(c) completing the process of filling dart before the operation begins — dart filling should never be done on moving vehicle or on elephant back,

(d) keeping multiple backup darts ready in case they are needed,

(e) use of spill-proof containers to transport drugs, and

(f) keeping human antidote ready before starting to fill the dart.

5. Animal-related risks: The wildlife being restrained — and also other animals in the forest — may attack the team that has ventured into their area. This can result in severe trauma through bites, horns, weight of the animal, etc. There is a chance of misjudging the anaesthetic depth, and the animal that has been darted — while appearing limp — may attack when approached. Some drug-induced muscular responses of the immobilised animal — such as sudden movements or bites — may also lead to accidents. Thus several precautions and preparations are warranted, including

(a) using pepper spray and loud horns to repulse away non-target animals,

(b) taking adequate safety measures to confirm the level of anaesthetic depth — such as using a long stick, a bamboo pole or pebbles to prod the seemingly anaesthetised animal — before approaching it,

(c) moving in forests only in vehicles or on elephant back, and never on foot,

(d) avoiding putting fingers between the teeth of the animal to reduce the risk emanating from involuntary muscular movements, and

(e) having sufficient backup of firearms.

Before starting the operation, one must always prepare an evacuation protocol and treatment options. If narcotics are to be used, the first aid kit must include naloxone, naltrexone, hydrocortisone, diazepam, atropine and adrenaline. In case of an emergency situation, the appropriate response may be remembered as HAD-ABC:

1. Help: Call for help

2. Absorption/antidote: Limit absorption, give antidote

3. Drip: Establish drip if indicated

4. Airway: Check airway and ensure that nothing is blocking it; remove fluids and tongue from the air passage and keep the head tilted to drain fluids

5. Breathing: Check breathing and give artificial respiration if required

6. Circulation: Check circulation and give CPR if required

7.6   CAPTURE MYOPATHY

Capture myopathy is a non-infectious metabolic disease of animals with significant morbidity and mortality, associated with pursuit, capture, restraint, and transportation of animals. The clinical signs include

1. stiffness in muscles,

2. severe pain in muscles,

3. ataxia — loss of full control of bodily movements,

4. paresis — muscular weakness with partial paralysis,

5. torticollis — turning of head to one side often due to spasms,

6. prostration — action of lying stretched on the ground,

7. paralysis,

8. animal typically becoming

   - obtunded — with dull sensitivity,

   - anorexic — with lost appetite, and

   - unresponsive, and

9. death — which can occur from within minutes or hours of capture to days or weeks after the inciting event.

It commonly affects herbivorous animals. The pathogenesis has three components:

1. perception of fear by the animal as it is being chased and captured,

2. activity of sympathetic nervous system and release of large amounts of adrenaline due to a 'flight or fight' response, and

3. intense muscular activity due to the action of adrenaline.

The pathophysiology can be described as:
Altered blood flow to the tissues due to effect of adrenaline → exhaustion of normal aerobic energy, particularly in skeletal muscles that are having intense muscular activity due to the action of adrenaline → exhaustion of ATP in muscle cells, together with decreased delivery of oxygen and nutrients, leading to an increased production of lactic acid, and inadequate removal of cellular waste products → damage to muscle cells, which makes them to undergo necrosis to varying degrees → myoglobin and creatinine kinase get released from these breaking cells and reach kidneys through blood, causing tubular necrosis in the kidneys and acute renal failure; similar necrosis of cardiac tissue can occur as well.

Thus we may say that capture myopathy is a very intense side-effect — including muscular rupture and failure of heart and kidneys — caused by release of large amounts of adrenaline and activity of sympathetic nervous system due to over-stimulation of the naturally evolved flight or fight response in animals when they perceive the fear of being captured.

The predisposing factors can be remembered with the mnemonic SECONDS:

1. S Species: mostly ungulates

2. E Environment: inclement temperature, rainfall, humidity, steep terrain, etc. can exacerbate the intensity of capture myopathy

3. C Capture related: speed of chasing (higher speed is more harmful), duration of restraint (prolonged restraint is more harmful), amount of handling (excessive handling generates more fear and is more harmful), positioning (unnatural positioning is more harmful), etc.

4. O Other diseases and co-morbidities in the animal: may intensify the impacts of adrenaline

5. N Nutrition: Vitamin E and selenium deficiency causes more harm

6. D Drugs: opioids and drugs that increase excitability add to the effect of adrenaline, while neuro-muscular blocking drugs which cause paralysis of the body without depressing the CNS generate intense stress and fear in the animal causing larger release of adrenaline

7. S Signalment: age (very young and very old are more affected), sex (mostly males affected), condition (pregnancy increases risk), etc.

Capture myopathy is also known by several other names that signify different facets of the disease:

1. muscular dystrophy: a disease that causes wasting of muscles

2. white muscle disease: a disease that turns muscles 'white' due to release of myoglobins

3. overstraining disease: a disease caused by overstraining the animal

4. capture disease: a disease involved with capturing of animal

5. cramp: a disease caused by excessive fatigue and involving painful involuntary contractions of muscles

6. leg paralysis

7. spastic paresis: spastic = relating to or affected by muscle spasm, paresis = muscular weakness with partial paralysis

8. stress myopathy: a disease of muscles caused by stress

9. transport myopathy: a disease of muscles seen during transport of animals

10. incipient myopathy: a disease of muscles that is incipient (beginning to happen or develop)

11. degenerative polymyopathy: a disease of multiple muscles that is characterised by progressive deterioration and loss of function in organs or tissues

12. muscle necrosis: death of muscles

13. idiopathic muscle necrosis: death of muscles that arises spontaneously and without a clearly discernible cause

During a post-mortem examination, capture myopathy presents itself as pale, ruptured muscles and presence of red fluid — due to myoglobinuria — in the urinary bladder.

The treatment of capture myopathy has a very low success rate. Treatment options include supportive treatments including

1. analgesics to reduce pain,

2. muscle relaxants to reduce spasm,

3. Vitamin E and selenium to act as anti-oxidants,

4. hyperbaric oxygen to reduce the effect of lactic acidosis,

5. sodium bicarbonate to reduce acidemia,

6. fluid therapy with balanced electrolyte solutions — often given intravenously,

7. intravenous nutritional support, and

8. muscle support and exercise.

In these conditions, prevention of capture myopathy is often the best approach for the animal. Prevention of capture myopathy requires several considerations, including

1. recognising the condition and precipitating factors of capture myopathy,

2. minimising handling of animals,

3. avoiding transportation or keeping it as brief as possible,

4. choosing drugs with rapid induction, rapid recovery, efficient delivery, and physiologic stability,

5. keeping anaesthesia as brief as possible,

6. use of tranquillisers to calm the animal,

7. efficient capture techniques, and

8. herd capturing which reduces fear and stress by keeping the herd together, e.g. the boma technique [Fig. 7.6].

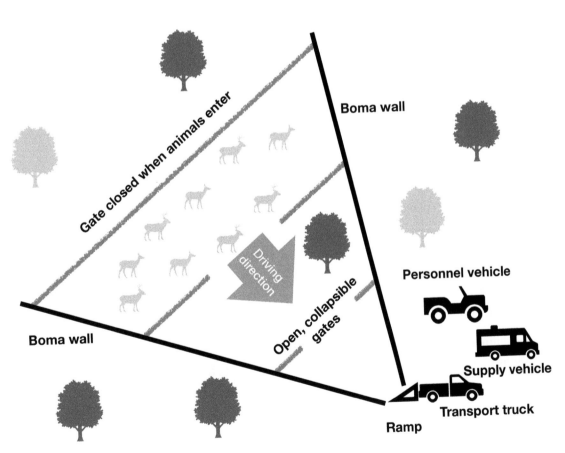

Figure 7.6: A representation of the boma technique. In this method, the herd is driven to a funnel-shaped structure with collapsible gates. As the herd moves into the narrower sections of the funnel, the gates are progressively closed. At the end of the funnel is a ramp through which the herd moves — *in toto* — into a transport truck and taken to the desired location to be swiftly released.

7.7 TRANSPORTING CAPTIVE WILD ANIMALS

When wild animals are to be transported — say to a site of release for genetic rescue — several points of care must be observed to enhance safety and success of the operation, and to reduce hardship to the animals. Some such points are:

1. Animals should be in good health. Very young and very old animals should be avoided and sub-adult animals preferred for transportation.

2. Pregnant and lactating females should be avoided for transport.

3. Antlered animals in velvet should be avoided since velvet is a soft tissue and any damage may lead to profuse bleeding.

4. Complete health check up — of animals, their handlers, and the transportation staff — should be done before transport.

5. Documentation regarding the animal, such as animal history cards, treatment cards and health certificates, and legal papers permitting transport of wild animals should accompany the animal during transportation.

6. Quarantine arrangements — if needed at the recipient station — should be ensured and inspected in advance so that the animals may be released as soon as possible.

7. The transport crates/containers should be of standardised dimensions to provide comfort to the animal without permitting excessive movement or somersaulting.

8. The floor of the transport crate/container should permit the animal(s) to stand comfortably and safely while removing all liquids that may get spilled on the floor. In particular, it should be ensured that the slats do not trap the feet of the animal(s).

9. Before transport, a thorough inspection should be made to ensure that

    (a) there are no sharp projections such as nails and screws in the transport container,
    (b) the transport container has thoroughly been disinfected,
    (c) the transport container does not have any residual chemicals due to disinfection/paints/preservatives that can harm the animal(s),
    (d) the transport container has adequate ventilation and the ventilation ports are unblocked, and
    (e) the lifting handles/griper bars for lifting are not bent or broken.

10. The animals should have adequate food, water, and suitable bedding such as straw for long journeys. For shorter journeys, animals should be fed before transportation.

11. During transportation, animals should be accompanied by a veterinarian and keepers to manage emergencies. This is especially essential when the animals are in a state of sedation. All necessary drugs, medicines, first aid kit, and restraining equipments should accompany the veterinarian, and additional ropes, buckets, water sprayers, and repair tools should accompany the keepers. Adequate provisions of money and the authority to spend it in unforeseen situations must be provided to some member of the transportation team.

12. The animals should be disturbed as little as possible during transport.

13. Very hot times (around noon), very cold times (after midnight) and extreme weather conditions should be avoided. If this is not possible, then controlled environment should be provided to the animals through use of air conditioners, water spray, heaters, etc.

14. With the exception of herds, birds and mothers with babies, animals should be kept singly.

15. Air lifting should be explored as an option to reduce time and duress to the animal. When resorted to, IATA/CITES guidelines should be followed for air transport.

16. The facilities such as zoos and veterinary hospitals on the way should be kept informed in advance about the movement of animals and their assistance should be solicited if needed.

CHAPTER 8

Wildlife genetics

Cats beget cats. Dogs beget dogs. When two Doberman Pinschers are mated, the puppies have the characteristics of Doberman Pinschers — long muzzle, shiny coat, long tail and soft ears. The puppies do not look like a Pomeranian or an Akita. Why is that so? Often when a child is born, we hear people say things like "It has mother's eyes" or "It has grandpa's nose," indicating that the characteristics of the previous generations make way into a newborn. How do the characteristics of parents get to express themselves in their offsprings? While it was a mystery for quite some time, we now know that these characteristics move through *genes*, and we study them in the discipline of *Genetics*.

8.1 GENES AND GENETIC DISORDERS

8.1.1 Genetics

Genetics is the study of heredity and the variation of inherited characteristics. It asks questions such as:

1. If the father has blood group AB, and the mother has blood group O, what will be the blood group of the child?

2. If the father is short and the mother is tall, will the child be tall, short or of an average height?

3. How can we breed plants with a desired set of traits — drought tolerance, resistance to insects and good productivity?

The science of Genetics is based upon the concept of gene, defined as "a unit of heredity which is transferred from a parent to offspring and is held to determine some characteristic of the offspring." We have genes for a variety of traits such as hair colour, tallness, and disease resistance. These genes move as discrete entities across generations, but may also change through *mutations*. Chemically, they are distinct sequences of nucleotides forming part of a chromosome. The order of the nucleotides holds information, and it often codes for a molecule that has a function. When the code changes, as through mutation, the genes change as well.

In chemistry, a nucleoside is a glycosylamine — a compound with a sugar (e.g. ribose) linked to an amine. The amine, depending on its structure, may be a purine — such as adenine or guanine, or a pyrimidine — such as thymine or cytosine. Nucleotide is a compound consisting of a nucleoside linked to a phosphate group. Nucleotides form the basic structural unit of nucleic acids such as DNA (Deoxyribonucleic acid) and RNA (Ribonucleic acid).

DNA resides in chromosomes. Chromosomes (from Greek *chroma* = colour and *soma* = body; referring to the coloured bodies observed in nuclei after staining procedures) are thread-like structures of nucleic acids and protein found in the nucleus of most living cells. DNA in the chromosomes carries the genetic information in the form of genes. We can use fluorescent stains (such as DAPI that binds to AT regions of DNA) to help visualise DNA, chromosomes or nuclei [Fig. 8.1].

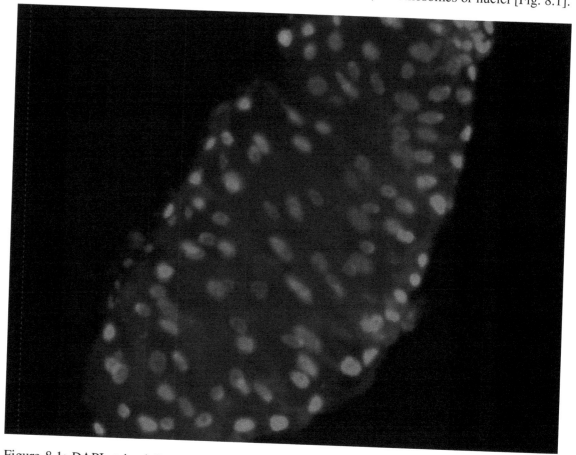

Figure 8.1: DAPI-stained *Drosophila melanogaster* salivary glands showing fluorescence where the stain binds to the AT regions of the DNA. Since DNA is in the chromosome which resides in the nuclei of cells, the nuclei of cells become clearly visible through fluorescence.

Since the information in a gene is encoded in the form of sequence of nucleotides, a change in this sequence will create variations in the gene. We define allele as each of two or more alternative forms of a gene that arise by mutation and are found at the same place on a chromosome. Since they are alternative forms of the same gene, they code for the same trait — defined as a *genetically determined characteristic* caused due to the presence of some allele. Good examples of traits are as flower colour and skin colour. For instance, a pea plant can have purple or white flowers. These variations represent the flower colour trait, and these are coded by alleles. The alleles written as P and p. P codes for purple flowers, and p codes for white flowers. Since genes exist in two copies, one coming from either parent, we can have three combinations of these two alleles — PP, pp and Pp. Alleles often have dominant-recessive relationships such that if an organism has two different alleles of the gene — Pp — the dominant allele expresses itself in the phenotype (observable trait), and the recessive allele does not find expression in the phenotype. The dominant allele is often written with a capital letter — such as P, and the recessive allele is often written as a small letter — such as p. Thus, an organism with Pp genotype (genetic constitution) will have purple flowers, which is the expression of the dominant allele P. PP will express as purple flowers and pp will express as white flowers.

Changes in the sequence of nucleotides are called mutations. They may result in the formation of new alleles of a gene, or even a new gene. Mutations may occur through the alteration of single or multiple base units in DNA. They may also occur through deletion, insertion, or rearrangement of sections of genes or chromosomes. Such changes usually occur during cell division as errors in copying of DNA sequences. They may also occur due to certain viruses. When mutations occur in the cells of the body (called *somatic mutations*), they result in non-heritable changes, though the organism becomes a *chimera* — an organism with cells with more than one distinct genotype. When mutations occur in the gametic cells (which form sperms or ova), the changes are heritable — they can get transmitted to subsequent generations. Such mutations are called *germline mutations*. Heritable changes play important roles in the evolution of species when the new traits that increase fitness get selected through natural selection [Ref: Section 8.2.1]. Some traits observed in wild type and mutant *Drosophila melanogaster* fruit flies are depicted in figure 8.2.

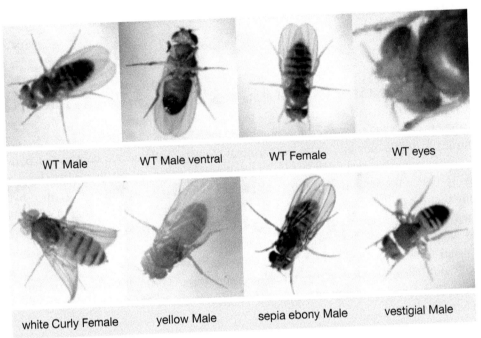

Figure 8.2: Some traits observed in wild type and mutant *Drosophila melanogaster*. The wild type fruit-flies have red eye colour, brown body colour, straight and long wings. *white* mutants have white eyes, *sepia* mutants have sepia-coloured eyes. *yellow* mutants have a yellow body colour, *ebony* mutants have a dark body colour. *Curly* mutants have curled wings, and *vestigial* mutants have short wings.

The different alleles of different genes make up the genetic constitution of an organism — called the *genotype*. The genotype is made visible through the expression of these genes. It is here that the environment also begins to play a role. This can be explained through the example of tallness. Suppose the gene for tallness has two alleles — T coding for tallness and t coding for shortness. Now consider an organism with a genotype TT. In normal conditions, this organism will be tall. But suppose it gets born during a time of severe drought and food scarcity. To grow tall, the organism needs ample food, especially lots of proteins. But due to the drought, the environment is unable to provide sufficient nutrition. In such conditions, even though the genotype predicts tallness, the organism may actually turn out to be very short — due to the environment.

We define *phenotype* as "the set of observable characteristics that arise through the interaction of the genotype with the environment." Thus, purple flowers are the phenotype produced by a PP genotype in pea plants.

In this chapter, we shall make extensive references to organisms such as the pea plant (*Pisum sativum*) and the fruit fly (*Drosophila melanogaster*). These are model organisms — species that are extensively studied to understand Biology and genetics. Often model organisms are those that are easy (and cost-effective) to rear, handle, manipulate, and observe. Since the principles of genetics apply to all organisms, the discoveries made in a model organism are applicable to other organisms as well.

The knowledge of genetics is crucial for the purpose of conservation since many genetic disorders manifest themselves in smaller, inbred populations. Conservation Genetics is the branch of science that aims to understand the dynamics of genes in populations, principally to avoid extinction, and includes topics such as cell line cryobanking in frozen zoos, chromosomal and karyotype analyses, population genetics and management, DNA sequencing and barcoding, genetic rescue and translocation of populations, hybrid and parentage identification, phylogenetics and phylogeography. We shall cover the basics of Conservation Genetics in this chapter, and the reader is encouraged to explore suitable references for a more detailed understanding. We begin with Mendel's laws of genetics.

8.1.2 Mendel's laws of genetics

The Austrian scientist and abbot, Gregor Johann Mendel, is recognised as the father of Genetics. He conducted some of the earliest experiments in this field, using pea plants as his model organism. The experiments consisted of generating 'purebred' lines — pea plants with certain traits (tall vs. dwarf, purple vs. white flowers, axial vs. terminal flowers, inflated vs. constricted pods, green vs. yellow pods, round vs. wrinkled seeds, and yellow vs. green seeds) that, when bred amongst themselves, produce the same traits in their offsprings. These purebred lines were then hybridised — meaning that the pollen from one plant was collected and dusted on the stigma of emasculated flowers (with male parts removed) of another plant — to create cross-fertilisation. The seeds that developed were then sown and tended till maturity. The traits in the offspring generation were noted and statistically analysed. At times, the traits were observed across several generations. At other times, multiple traits were observed at the same time. Through such experiments, Mendel discovered the fundamental tenets of genetics, now referred to as Mendel's laws of genetics in his honour. These laws are:

1. The law of dominance: Recessive alleles are masked by dominant alleles. Before the discovery of this law, the common understanding was that the offsprings will have characteristics that are in between those of their parents. For instance, if one parent is tall and the other parent is short, the offsprings will be of an average height. Similarly, in the case of pea plants, if one parent produces purple flowers and the other parent produces white flowers, the then contemporary reasoning predicted that their offsprings should have flowers with a shade in between purple and white. Mendel tested this hypothesis through experimentation. He found that when pure-bred plants with purple flowers (PP) are crossed with pure-bred plants with white flowers (pp), the offsprings (Pp) always produce purple flowers — never white, and never a shade between purple and white — meaning that the trait of the recessive allele ($p \implies white$) gets completely masked by the trait of the dominant allele ($P \implies purple$). He observed this law over several traits:

Characteristic	Dominant trait	Recessive trait
Stem height	Tall	dwarf
Flower colour	Purple	white
Flower position	Axial	terminal
Pod shape	Inflated	constricted
Pod colour	Green	yellow
Seed shape	Round	wrinkled
Seed colour	Yellow	green

2. The law of segregation: The two alleles of a gene separate (segregate) during gamete formation — a pollen or an ovum carries only one allele of each pair. For example, a *Pp* plant will
produce pollen and ova that have either *P* or *p*, but never both together.

With these two laws, we can explain the results of monohybrid crosses — those in which
only one characteristic is studied at a time. For example, Mendel had observed that when
pure-bred purple flower producing plants [Parent generation] were crossed with pure-bred
purple white producing plants [Parent generation], all the resulting offsprings [called the
first filial generation, F1] produced purple flowers. However, when the plants in the first
filial generation were crossed amongst themselves, their offsprings [called the second filial
generation, F2] produced either purple flowers or white flowers. This meant that the trait
of producing white flowers was not lost in the first cross — it was merely masked by the
trait of producing purple flowers. Mendel also noted that in the second filial generation, the
plants producing purple flowers and the plants producing white flowers were always found
in the same ratio — 3:1. This could only be explained if the alleles were segregated during
each cross. We may represent these crosses using Punnett square diagrams. In a Punnett
square diagram, the gametes from father are represented on the top and the gametes from
the mother are represented on the left. The genotype of the offsprings that would result
from these gametes are written in each cell of the table, permitting an easy understanding of
genotypic and phenotypic ratios.

For the example case: *PP* × *pp*, the Punnett square will be written as:

F1 for *PP* × *pp*

	p	*p*
P	*Pp*	*Pp*
P	*Pp*	*Pp*

Here we observe that all the offsprings have the same genotype — *Pp*. Since this genotype
produces purple flowers (as *P* is dominant over *p*), all the offsprings in the F1 generation will
produce purple flowers.

When the individuals in the first filial generation are crossed amongst themselves: *Pp* × *Pp*,
the Punnett square will be written as:

F2 for *Pp* × *Pp*

	P	*p*
P	*PP*	*Pp*
p	*Pp*	*pp*

Here we observe that the offsprings can have three kinds of genotypes: *PP*, *Pp* and *pp*. Out
of four offsprings, one is *PP*, two are *Pp* and one is *pp*. Thus the genotypic ratio can be
written as *PP* (1) : *Pp* (2) : *pp* (1), or 1:2:1.

We can similarly compute the phenotypic ratio. Since *PP* and *Pp* genotypes produce purple
flowered phenotype and *pp* genotype produces white flowered phenotype, out of the four
offsprings, three will produce purple flowers and one will produce white flowers. Thus the
phenotypic ratio can be written as Purple flowered (3): white flowered (1), or 3:1.

Thus, the law of segregation explains the 3:1 F2 ratio. It is important to note here that in genetics, we are concerned with probabilities. If we consider 4 offsprings in the F2 generation,

it is not always necessary that 3 will produce purple flowers and 1 will produce white flowers. It is even possible that just out of chance, all the offsprings produce white flowers. But when we consider a very large number of observations, we will have the 3:1 ratio. Thus, when we consider 4,000 offsprings in the F2 generation, we will have *around* 3,000 individuals that produce purple flowers and *around* 1,000 individuals that produce white flowers. The ratio may be something like 2,960:1,040 or 3,055:945, but will always be near 3:1.

3. The law of independent assortment: Each pair of alleles separates into the gametes independently of other pairs. Thus if we consider parents with yellow and round seeds ($YYRR$ genotype) and with green and wrinkled seeds ($yyrr$ genotype), the alleles of seed colour and seed shape will separate independently of each other. This means that YR and yr will not move together in the gametes. This law was derived from the observation that in dihybrid crosses — those in which two traits are studied simultaneously — the F2 generation always depicted the 9:3:3:1 phenotypic ratio.

For the example case: $YYRR \times yyrr$, the Punnett square will be written as:

F1 for $YYRR \times yyrr$

	yr	yr
YR	$YyRr$	$YyRr$
YR	$YyRr$	$YyRr$

We observe that all the offsprings have the same genotype — $YyRr$. This genotype produces yellow and round seeds (as Y is dominant over y and R is dominant over r). Hence, all the offsprings in the F1 generation will produce yellow and round seeds.

When the individuals in the first filial generation are crossed amongst themselves: $YyRr \times YyRr$, the Punnett square will be written as:

F2 for $YyRr \times YyRr$

	YR	Yr	yR	yr
YR	$YYRR$	$YYRr$	$YyRR$	$YyRr$
Yr	$YYRr$	$YYrr$	$YyRr$	$Yyrr$
yR	$YyRR$	$YyRr$	$yyRR$	$yyRr$
yr	$YyRr$	$Yyrr$	$yyRr$	$yyrr$

Thus in F2 we have a variety of genotypes. The genotypic ratio can be written as $YYRR$ (1) : $YyRR$ (2) : $YYRr$ (2) : $YyRr$ (4) : $yyRr$ (2) : $Yyrr$ (2) : $YYrr$ (1) : $yyRR$ (1) : $yyrr$ (1).

The phenotypic ratio can be written as Purple flowered, Green pod (9) : white flowered, Green pod (3) : Purple flowered, yellow pod (3) : white flowered, yellow pod (1), or 9:3:3:1.

Thus the law of independent assortment explains the 9:3:3:1 F2 ratio.

8.1.3 Variations to complete dominance

Biology is a science of exceptions, and we do find numerous instances where Mendelian Genetics doesn't completely apply. Some prominent examples are:

1. Codominance: Codominance occurs when contributions of both the alleles are visible in the phenotype. A good example is the human blood group, determined by alleles I^A, I^B and I^O.

Since these alleles determine the antigen(s) on the surface of red blood cells, we can have multiple scenarios:

Genotype	Surface antigens	Blood group
$I^O I^O$	None	O
$I^O I^A$	A	A
$I^O I^B$	B	B
$I^A I^A$	A	A
$I^A I^B$	A and B	AB
$I^B I^B$	B	B

2. Incomplete dominance: Incomplete dominance occurs when the phenotype of the heterozygous genotype is distinct from, and often intermediate to the phenotypes of the homozygous genotypes. A good example is flower colouration in common snapdragon (*Antirrhinum majus*) — the alleles R (for red colouration) and r (for white colouration) show incomplete dominance. Thus, while RR plants bear red flowers and rr plants bear white flowers, Rr plants bear pink flowers — intermediate between red and white.

3. Sex-linked traits: Those alleles that are present on the sex chromosomes (X and Y) will only move with those sex chromosomes. Thus, sex-specific patterns of inheritance and presentation are observed when a gene present on a sex chromosome mutates. A good example is the eye colouration gene present on the X chromosome of fruit fly *Drosophila melanogaster*. The X chromosome may have the wild type gene, represented as X^{w+}, with the '+' indicating the typical phenotype of red eye colouration. Else the X chromosome may bear the mutant of this gene, represented as X^w. The Y chromosome does not bear the eye colouration gene.

In this case, when homozygous red-eyed females are crossed with white-eyed males, we may write the following Punnett square:

F1 for $X^{w+}X^{w+} \times X^w Y$

	X^w	Y
X^{w+}	$X^w X^{w+}$	$X^{w+} Y$
X^{w+}	$X^w X^{w+}$	$X^{w+} Y$

In this case, the genotypic ratio is $X^w X^{w+}$ (2) : $X^{w+} Y$ (2), or 1:1.

Phenotypically, $X^w X^{w+}$ individuals are red-eyed females (since two X chromosomes make females, and X^{w+} is dominant over X^w). Similarly, all $X^{w+} Y$ individuals are red-eyed males (since the Y chromosome makes males, and they only have one wild type eye colouration allele on the X chromosome).

On the other hand, when homozygous white-eyed females are crossed with red-eyed males, we may write the following Punnett square:

F1 for $X^w X^w \times X^{w+} Y$

	X^{w+}	Y
X^w	$X^{w+} X^w$	$X^w Y$
X^w	$X^{w+} X^w$	$X^w Y$

In this case, the genotypic ratio is $X^{w+} X^w$ (2) : $X^w Y$ (2), or 1:1.

Phenotypically, $X^{w+} X^w$ individuals are red-eyed females (since two X chromosomes make females, and X^{w+} is dominant over X^w). Similarly, all $X^w Y$ individuals are white-eyed males (since the Y chromosome makes males, and they only have one mutant eye colouration allele on the X chromosome).

Similarly, when heterozygous red-eyed females are crossed with white-eyed males, we may write the following Punnett square:

F1 for $X^w X^{w+} \times X^w Y$

	X^w	Y
X^w	$X^w X^w$	$X^w Y$
X^{w+}	$X^w X^{w+}$	$X^{w+} Y$

In this case, the genotypic ratio is $X^w X^w$ (1) : $X^w Y$ (1) : $X^w X^{w+}$ (1) : $X^{w+} Y$ (1), or 1:1:1:1.

Phenotypically, $X^{w+} X^w$ individuals are red-eyed females (since two X chromosomes make females, and X^{w+} is dominant over X^w) and $X^w X^w$ individuals are white-eyed females. Similarly, $X^w Y$ individuals are white-eyed males (since the Y chromosome makes males, and they only have one mutant eye colouration allele on the X chromosome) and $X^{w+} Y$ individuals are red-eyed males (since the Y chromosome makes males, and they only have one wild type eye colouration allele on the X chromosome).

Such seemingly bizarre ratios indicated that the gene for eye colouration must be on the X chromosome.

8.1.4 Chromosomal disorders

DNA resides in the chromosomes. Thus, chromosomal disorders become important for an organism's health. Chromosomal disorders may manifest in the form of missing, extra, or irregular portion(s) of chromosomal DNA, and are often caused by an atypical number of chromosomes (more or less than normal) or structural abnormalities in one or more chromosomes. An individual with the appropriate number of chromosomes for their species is known as a "euploid individual." An individual with an error in the number of chromosomes, or deletion or duplication of portions of chromosomes is known as an "aneuploid individual."

Many things can go wrong with chromosomes, and so we have several kinds of chromosomal disorders, such as

1. Numerical disorders — abnormalities in the number of chromosomes found in the cells, including

    (a) monosomy — just one chromosome in the place of a usual pair of chromosomes,

    (b) trisomy — three chromosomes in the place of a usual pair of chromosomes,

    (c) tetrasomy — four chromosomes in the place of a usual pair of chromosomes, etc.

2. Structural abnormalities — abnormalities in the structure of chromosomes, including

    (a) deletion, where a part of a chromosome is missing,

    (b) duplication, where a part of a chromosome is present in two or multiple copies,

    (c) translocation, where a part of a chromosome gets shifted to another chromosome, and

(d) inversion, where a part of a chromosome is turned upside down. Inversions are classified as

    i. pericentric (Greek *peri* = around) inversions — inversions involving the centromere, and

    ii. paracentric (Greek *para* = next to) inversions — inversions not involving the centromere.

etc.

8.1.5 Genetic disorders

On a finer scale, we have genetic disorders — problems caused by abnormalities in the genome. They could include situations

1. where a gene does not work, say, due to deletion or inactivation, or

2. where a gene works extra, say, due to duplication or extra activation, or

3. where a gene works differently, say, due to a mutation that changes the structure of the protein that is coded by the gene.

The working of genetic disorders can be explained by the Central dogma of Molecular Biology — information flows from DNA, through RNA to proteins:

$$\text{DNA} \xrightarrow{\text{Transcription}} \text{RNA} \xrightarrow{\text{Translation}} \text{Protein}$$

Thus, if there are changes in the DNA (caused by mutation(s) in a gene), the protein that gets made will have an incorrect information (in the form of aberration(s) in the sequence of amino acids). Since the sequence of amino acids determines the structure of the protein, we may have a situation where the protein does not work, works excessively, or works differently than normal. And this may lead to certain genetic disease(s).

It is important to note here that just like genes, the abnormalities in the genes get passed from one generation to the next through heredity. In the context of wildlife management, it is crucial that these abnormalities are detected and managed. The management may be in the form of removal of individuals that have genetic disorders — so that the mutation does not spread in the population — or in the form of creation of conditions such that the mutations are unable to express themselves and new mutations do not arise.

Many mutations are recessive, and only get to express themselves when present in a homozygous condition. Since homozygosity can easily arise due to mating of close relatives, the management of inbreeding depression [Ref: Section 8.4] — especially those occurring in small populations — becomes crucial for the conservation of wildlife. It is also essential to understand genetics at the scale of populations — how do mutations spread in a population, how do they express themselves in a population, and why has nature permitted a situation where errors occur and spread (Hint: without errors there will not be variations — and biodiversity!). In the next section, we discuss population genetics — genetics at the level of populations.

8.2 POPULATION GENETICS

A population is defined as "a localised group of individuals that are capable of interbreeding and producing fertile offsprings." This 'localised group' is not static — it keeps on changing. The size may increase or decrease, the proportion of animals in different age classes may change — making the population younger or older, the sex ratios may change, or there may be changes at the level of genes. This continuous change is what endows a population with *resilience* and an ability to adapt to its ever-changing environment. For instance, in times of food scarcity, the population size may shrink. With less mouths to feed, the chance of at least some individuals surviving increases, for not everybody will die of starvation. The death of large number of individuals allows the population to concentrate its resources to few individuals — usually, those that are sturdier, dexterous, or otherwise better suited to overcome the shortage. And when later a time comes when there is ample food available, these few surviving individuals will breed and increase the size of the population, permitting the population to make the maximum utilisation of the available resources. If the population was static — and with no differences between individuals, all the individuals would have perished during times of scarcity. And in times of abundance of resources, the static population wouldn't have done an efficient utilisation of resources either.

The environment may change in several other complex ways as well, requiring changes in the gene pool. A good example is the evolution of the peppered moth (*Biston betularia*). The peppered moth is an insect found in temperate regions. The insect exists in two forms — a light form (called morpha *typica*), and a dark form (called morpha *carbonaria*). Before the industrial revolution, almost all the specimen of the peppered moth were of the *typica* form. This is because when these moths rested against the trees, their lighter colours provided them a good camouflage — it was difficult to discern a light-coloured moth resting on a light-coloured tree. On the other hand, a dark-coloured moth resting on a light-coloured tree was very conspicuous, and could easily be spotted by insectivorous birds. Thus, the dark-coloured moths were preferentially removed from the population, with the result that the majority of the peppered moths were of a light-coloured variety.

Then came the industrial revolution. The smoke released from the factories started to kill off the lichen on the tree barks, with the result that the tree barks began to become darker. On these darker backgrounds, the light-coloured forms could be easily spotted by the birds, but the dark-coloured moths were less conspicuous. As a result, the birds now ate the light-coloured moths in much larger numbers than the dark-coloured moths. And with the passage of time, the dark-coloured moths (morpha *carbonaria*) became the majority.

If there were no variations in the moth colours to begin with, the population would probably have collapsed during the industrial revolution. But the presence of the variations, and changes in the population at genetic levels (represented by the changed frequencies of the two alleles) saved the population from an imminent collapse due to a change in the environment.

The story has one more twist. In the twentieth century, as people became more and more concerned with the deplorable quality of air around them, air laws were passed to limit the amount of smoke that could be emanated by the factories. As the air became cleaner, the lichen grew back. And with the backgrounds changing, the lighter *typica* form again got a survival advantage against the darker *carbonaria* form. As a result, the allele frequencies began to change once again, and now we find many more *typica* forms than the *carbonaria* forms!

Population genetics studies such changes — how the populations change genetically, over time. And the lessons from population genetics are of crucial importance to the conservation of wildlife.

8.2.1 Relationship with evolution

The science of Population Genetics has a close relation with the science of Evolution — the **genetic adaptation** *of organisms to their environment*. Genes change in their frequencies often as a response to changes in the environment, just as in the case of the peppered moths. Such changes permit the individuals in the population to adapt to their surroundings. This means that they become better able to survive and multiply in their environment.

We have seen before — in our discussion on the *Central dogma of Biology* — that information passes from DNA to proteins, mediated by RNA. In the context of peppered moths, changes in their colouration required changes in proteins — to create more and more pigments that make a moth darker. This required changes in the information contained in the DNA — meaning changes in the genes. This is what evolution is all about — changes in the genes that permit the organisms to adapt to their environment. Since genes are inherited, we may also say that genetic adaptation is "inheritable fitness."

Fitness is the ability of a particular organism to leave more descendants in the future generations, relative to other organisms. When the tree trunks were light in colour, the *typica* forms were 'fitter' than the *carbonaria* forms. This is because the *typica* forms were better protected, and thus had a better ability to survive to the breeding stage, giving rise to offsprings that had their genes for lighter colouration. On the other hand, since the *carbonaria* forms were more conspicuous, they were eaten by the birds and so each generation had very few number of individuals born out of *carbonaria* parents. With the change in environmental conditions, the fitness of these forms also changed.

The process of evolution works to maximise fitness of individuals through the process of natural selection. That is, those individuals that have greater fitness, get 'selected' naturally, with the result that with time, the 'fitness' of the species increases. But it is also possible that these changes make the species so different from their earlier versions that they are now no longer the same species. This is how new species come into existence — each fitter than the earlier species in a particular environment. Thus, to understand evolution, we must understand fitness.

Fitness has certain important characteristics:

1. Fitness is environment-specific. If the environment changes, the fitness may change. We have observed a good example in the form of industrial melanism in peppered moth — when the environment changed (became more or less polluted), the fitness of *typica* and *carbonaria* forms changed.

2. Fitness is species-specific. When the environment warms due to global warming, the fitness of species adapted to warmer conditions may increase, while the fitness of species of colder climates will reduce. This plays an important role in the phenomenon of 'escalator to extinction,' which was discussed in Chapter 4.

3. High reproductive rate alone does not mean higher fitness; but higher survival of more progeny does. This is because for traits to be continuously selected, they must be passed on to subsequent generations. If most of the progeny die before reaching maturity and reproducing themselves, the genes of the parent will be lost. In such cases, we'll say that the parents have low fitness.

4. Fitness should be measured across several generations — it is a long-term measure. Evolution itself acts slowly, over several generations.

5. Fitness works at the level of a complete organism, not on individual traits such as size or speed. The largest or the fastest organism may not be the most efficient when it comes to foraging, eluding its predators, fighting diseases, or getting a mate — prerequisites to surviving to maturity and producing offsprings. This explains why most organisms are 'average' — the 'averageness' comes from optimal resource allocation to several traits, all of which together result in a higher fitness of the organism.

In the process of natural selection, only those organisms that are best adapted to their environment tend to survive and transmit their genetic characteristics to the succeeding generations; while those less adapted tend to be eliminated. Thus, over several generations, the adaptability to the environment increases and the organisms become more and more specialised. Natural selection happens in five steps:

1. Variation: All individuals are not identical; they have different characteristics. If everyone was the same (a clone of each other), there would be nothing to select for or against. We can observe variations in several characteristics of organisms — size, speed, agility, keenness of senses, aggression, etc. It is upon this diversity of characteristics that the process of natural selection works. This also explains why a certain rate of mutations is permitted and has been selected by the process of natural selection. If there were no mutations, the variety of characteristics would reduce, with important ramifications to natural selection and thus the ability of the species to survive changes in the environment.

2. Overpopulation: Organisms tend to produce excess offsprings — e.g. female mosquitos may lay 500 to 1,000 eggs. This excess of individuals creates the raw material on which natural selection can act. Only when there are several individuals can a few be selected and others eliminated.

3. Struggle for existence: Resources are limited, so not all offsprings will be accommodated. This is due to the limited carrying capacity of the environment. Out of the 500 to 1,000 eggs that are laid by the female mosquito, only around 2 will survive to maturity when the environment is near the carrying capacity. Others will die — probably due to scarcity of food, or due to being eaten by predators such as frogs and insectivorous birds. So all of the 500 to 1,000 individuals that are born will have to struggle against each other — a struggle for existence and survival.

4. Survival of the fittest: Only those individuals best able to obtain and use resources will survive and reproduce. This is because they will 'out-compete' others — say by being most efficient in getting food and dodging predators. When these best-fit individuals survive, they carry with them the characteristics that made them best fit. These characteristics will then be passed on to the next generation through heredity.

5. Changes in the gene pool: Inherited characters increase the frequency of favoured traits in the population. When the environment is full of predators, the ability to survive the predators is a favoured characteristic. This can be achieved, say, through stealth (better camouflage) or by flying fast (better muscles). Because only those individuals will survive that have better camouflage and can fly faster, when they reproduce, these traits will get passed on to the next generation. In other words, the alleles that result in these traits will be passed on to the next generation, and the alleles that result in sub-optimal traits (such as poor camouflage and slow speed) will not be passed to the next generation. In this way, the frequency of the alleles in the gene pool will change — and more individuals in the population will have the alleles that result in the best-suited traits.

Thus, natural selection acts by changing the allele frequencies in the gene pool through survival and reproduction of those individuals that are best suited to the environment.

Over time, this may result in the creation of new species. Consider a population frugivorous birds living on an island. There is plenty of food available on the island, and the birds feed on fleshy fruits of trees and hard fruits of shrubs. Now consider changes in the environment. On one portion of the island, there is a forest fire, and the shrubs with all their seed banks get killed in the forest fire. On another portion, there are repeated droughts killing off the trees. Thus the birds face a shortage of food on both portions of the island. The struggle for existence increases, creating a greater pressure of natural selection. In the first portion, those birds that are better suited to get their food from the fleshy fruits of the trees will be fitter. Perhaps eating the fleshy fruits requires a flexible and sharp beak. Then the process of natural selection will select those individuals that have a flexible and sharp beak, and after several generations, we will find that most of the birds on this portion of the island have flexible and sharp beaks. In the second portion, those birds that are better suited to get their food from the hard fruits of the shrubs will be fitter. Perhaps eating the hard fruits requires a hard but blunt beak. Then the process of natural selection will select those individuals that have a hard and blunt beak, and after several generations, we will find that most of the birds on this portion of the island have hard and blunt beaks. Thus, both the sub-populations have now become extremely specialised when it comes to their food.

In these situations, when these two sub-populations breed among each other, the progeny that arise will have characteristics from both the parents — perhaps in the form of flexible (if the allele for a flexible beak is dominant over the allele for a hard beak) and blunt (if the allele for a blunt beak is dominant over the allele for a sharp beak) beaks. Now these individuals will be less fit in eating the fleshy fruits (because they do not have sharp beaks), and also will be less fit in eating the hard fruits (since they do not have hard beaks). Thus, in whichever portion of the island they go to, they will be out-competed by the pure bred residents. In other words, these individuals will not be selected by the process of natural selection, since they are not fitter than the pure-breds.

Over several generations, the two sub-populations will continue to diverge from each other, since there is no interbreeding among them. And in the course of time, they will become so different that they no longer are able to interbreed — perhaps due to changes in their anatomy or preferred breeding seasons. In these situations, they will have become two different species.

In this way, natural selection results in speciation — evolution of new species. Let us now examine this process from the perspective of population genetics.

8.2.2 Population variation and evolution

Population variation is "the distribution of phenotypes among individuals." It asks questions such as:

1. What is the height distribution in a population?

2. What is the distribution of skin colours in a population?

3. What is the distribution of weights in a population? etc.

Since phenotypes result from an interaction of the genotype with the environment, we may scrutinise population variation at the genetic level. We define gene pool as "the total genetic diversity in a population or a species at any time." That is, if we consider all the genes with their various alleles together, we get the gene pool. A large gene pool indicates that there are a large number of genes and many variations (alleles) in the population or the species. As we have seen before, this variety provides the raw material on which natural selection can act. The genetic variety also increases the resistance and resilience of the population in the face of environmental changes.

In the population, the proportion of an allele is called its allele frequency. For example, if we consider a population with 640 plants with red flowers (RR), 320 plants with pink flowers (Rr) and 40 plants with white flowers (rr) — showing an incomplete dominance phenotype, we can compute the number of alleles and allele frequencies as follows:

Number of R alleles = $640 \times 2 + 320 \times 1 + 40 \times 0 = 1280 + 320 = 1{,}600$

Number of r alleles = $640 \times 0 + 320 \times 1 + 40 \times 2 = 320 + 80 = 400$

Allele frequency of $R = \frac{1600}{1600+400} \times 100\% = 80\%$

Allele frequency of $r = \frac{400}{1600+400} \times 100\% = 20\%$

8.2.2.1 Hardy-Weinberg principle

Evolution results from a change in the allele and genotype frequencies, leading to more specialised (fitter) individuals, and over a long time, speciation. Thus, an absence of evolution should be characterised by no changes in the allele and genotype frequencies. This is the crux of the *Hardy-*

Weinberg principle, which states that in the absence of any evolutionary influences, the allele and genotype frequencies in a population shall remain constant from generation to generation.

Thus, in our example of the plant population, we may write:

Allele frequency of $R = 80\% \implies$ f(R), $p = 0.8$

Allele frequency of $r = 20\% \implies$ f(r), $q = 0.2$

In the absence of evolutionary influences, not only will p and q remain constant, but also the proportions of individuals. So,

Proportion of RR individuals $= 0.8 \times 0.8 = 0.64$

Proportion of rr individuals $= 0.2 \times 0.2 = 0.04$

Proportion of Rr individuals $= 1 - (0.64 + 0.04) = 0.32$

This can also be represented with a Punnett square for Hardy-Weinberg equilibrium:

Progeny for $Rr \times Rr$

	R (p)	r (q)
R (p)	$RR(p^2)$	$Rr(pq)$
r (q)	$Rr(pq)$	$rr(q^2)$

In this case,

$p + q = 1$, and

$p^2 + 2pq + q^2 = 1$

which should be, since

$(p+q)^2 = p^2 + 2pq + q^2$

Note that this is the same as the current proportions:

Proportion of RR individuals $= \frac{640}{640+40+320} = 0.64$

Proportion of rr individuals $= \frac{40}{640+40+320} = 0.04$

Proportion of Rr individuals $= \frac{320}{640+40+320} = 0.32$

In a population with several alleles $A_1, A_2, A_3 \dots A_n$ with allele frequencies $p_1, p_2, p_3 \dots p_n$, the Hardy-Weinberg principle can give a generalised equation:

$p_1 + p_2 + p_3 + \dots + p_n = 1$

In this case, the frequencies of homozygotes and heterozygotes will be given by the expansion of

$(p_1 + p_2 + p_3 + \dots + p_n)^2$

Thus, Hardy-Weinberg principle represents a population that is not evolving. If the population evolves, we must observe a violation in the Hardy-Weinberg principle. Or in other words, if we observe a violation in the Hardy-Weinberg principle — say through changing allele frequencies — then the population must be evolving.

8.2.2.2 *Evolution: Violations of Hardy-Weinberg equilibrium*

Evolution, or a change in the gene pool of the population may occur due to several reasons, each of which will violate the Hardy-Weinberg principle. These reasons can be:

1. Mutation — Mutation creates new alleles, those that were not present before. The creation of a new allele in the gene pool changes the allele and genotype frequencies in a population, and so is a violation of the Hardy-Weinberg principle. In some cases, the mutation is so drastic that it jumpstarts the process of speciation — say by providing a large survival advantage to the individuals that have the mutation. In other cases, it creates variations that will be acted upon by natural selection over several generations.

2. Migration (Gene flow) — When individuals from an outside population come into a population, they bring with them alleles that were either not present in the home population, or were present at different frequencies. This addition of new alleles or changes in the allele frequencies is a violation of the Hardy-Weinberg principle. In the case of plants, the migration of genes and alleles can occur through processes such as pollination [Fig. 8.3a] — the pollen from plants outside the home population represent the allele frequencies of the outside population, which may be different from the allele frequencies of the home population.

3. Small population effect — Random changes due to sampling may result in changes in the allele frequencies through a process called the genetic drift. This may occur even when there is no preferential mating — i.e. the mating is completely random [Fig. 8.3b], but only a small fraction of individuals are able to mate.

This can be explained by the analogy of marbles in a jar [Fig. 8.3c]. The different coloured marbles in the jar represent different alleles in a population. If only a small number of marbles are selected to form a new population, the new population may have an allele frequency that is very different from that of the original population.

In nature, such selections occur during population bottlenecks. A population bottleneck is a sharp reduction in the size of a population. This may occur due to catastrophes such as large-scale famines, droughts, floods, fires, and diseases, and also due to human activities such as large-scale poaching and habitat destruction. When such events occur, only a small number of individuals from the original population remain, changing the allele frequencies drastically.

A good example of population bottleneck is the Asiatic lion population found in the Gir National Park [Fig. 8.3d]. While the current population of Asiatic lions exceeds 600 individuals, at one time we only had a dozen individuals left out of several thousands — primarily due to excessive poaching during Mughal and British periods [Fig. 8.3e]. This represents a population bottleneck. When such few individuals were left, their genes and alleles formed the gene pool. Most of the breeding was possible only between close relatives. Today we can observe the impacts of population bottleneck and inbreeding in the surviving lions.

4. Non-random mating — For Hardy-Weinberg principle to operate, there should be a good 'mixing' of alleles in the population — through random mating. Instance of non-random mating — such as inbreeding and mate preference — result in changes to allele frequencies.

Inbreeding is the breeding between close relatives — parents and offsprings, brothers and sisters, close cousins, etc. This may occur when there are very few individuals left in a population, leaving inbreeding the only option. This is often exacerbated when other populations live very far off, or are separated by geographical discontinuities such as mountain ranges or oceans. In some species, inbreeding also occurs due to mate preference, where close relatives are preferred for mating.

Mate preference occurs when certain characteristics of the potential mates are preferred during courtship displays. A good example is mating in peacocks (Pavo cristatus) [Fig. 8.3f]. During their courtship displays, the male peacocks 'dance,' spreading their tail wings. The male peacocks with the largest, shiniest and most symmetrical tail wings are selected by the females for mating. This is strange, considering the fact that peacocks can only fly to small distances. When disturbed, they run, and rarely fly away. And the selection is done through the tail wings, not the flight wings. In fact, the long tail wings actually act as a hindrance to smooth flight! In essence, the female peafowls select handicapped males as their preferred potential mates!

But the fact that this behaviour of the female peafowls has been 'selected' through evolution means that it should be serving some purpose. The mate preference in peafowls can be explained in two ways:

(a) Handicap principle — Since the large tails of peacocks carry an extra risk of getting predated upon, so only the best males can survive this risk while still having large tails. If they were not otherwise 'fit,' the large-tailed males would already be dead! Thus, the presence of large tails indicates that these males have many desirable qualities — perhaps they are capable of fighting back, or of running very fast! And these desirable qualities will be passed on to the offsprings after mating.

(b) Good genes hypothesis — Large, shiny, symmetrical tails require lots of inputs to be constructed. Thus, their presence indicates that the males donning these large tails have a particularly efficient metabolism. They are also good at foraging, and have a good ability to fight off diseases. These characteristics — efficient metabolism, foraging adeptness and good health — are desirable qualities which will be passed on to the offsprings after mating.

Both handicap principle and good genes hypothesis are examples of honest signals — qualities that gives a true impression of an individual's fitness. Hence the mate preference for the long tails.

5. Selection — Natural selection can change allele frequencies when fit individuals survive and procreate, while the unfit individuals die. Such selection can be of three kinds:

(a) Directional selection, which occurs when one extreme phenotype is favoured over other phenotypes. A good example is the industrial melanism in peppered moth — during the darkening of trees, one form — *carbonaria* — was selected, while during the lightening of trees, the other extreme form — *typica* — was selected. Such selections, over time, result in shifting of allele frequencies in the direction of the selected phenotype.

(b) Disruptive selection, which occurs when two extreme values of a trait are favoured over the intermediate values. A good example is body colouration for camouflage in an environment with light sand and dark rocks — light-coloured individuals can hide in the backdrop of light sand, while dark-coloured individuals can hide in the backdrop of dark rocks. The individuals with intermediate body colours cannot hide anywhere — they will be conspicuous in both the backdrops. If the species is a prey species, it will easily be spotted by the predators — and then be eaten. If the species is a predator species, it will find approaching the prey difficult without a good camouflage. Thus the lack of camouflage will reduce the fitness of the individuals with intermediate body colours. In such situations, both the extreme values of the trait — here very dark and very light body colours — will get selected over the intermediate values.

(c) Stabilising selection, which occurs when the mean values of phenotypes are favoured over the extreme values. A good example is body colour for camouflage in a single colour environment such as a desert with brown sand. The individuals with brown body colour — resembling the sand colour — will be much better camouflaged than those with lighter or darker body colours. This will result in increased fitness of sand-brown individuals over lighter and darker individuals, leading to a stabilising selection.

The learnings from Population Genetics can be put to use to analyse the viability of populations — by computing the chances of events like inbreeding depression. This forms a part of the discipline of Population viability analysis.

(a) Pollination is one way by which genes can 'flow' across populations.

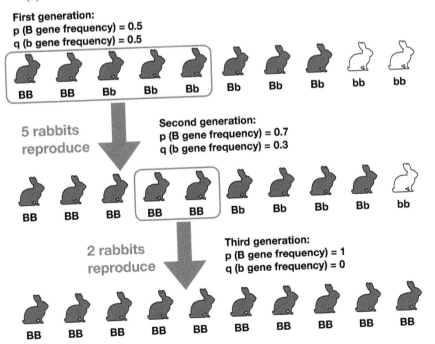

First generation:
p (B gene frequency) = 0.5
q (b gene frequency) = 0.5

BB BB Bb Bb Bb Bb Bb Bb bb bb

5 rabbits reproduce

Second generation:
p (B gene frequency) = 0.7
q (b gene frequency) = 0.3

BB BB BB BB BB Bb Bb Bb Bb bb

2 rabbits reproduce

Third generation:
p (B gene frequency) = 1
q (b gene frequency) = 0

BB BB BB BB BB BB BB BB BB BB

(b) Random mating may also result in altered allelic frequencies in a population.

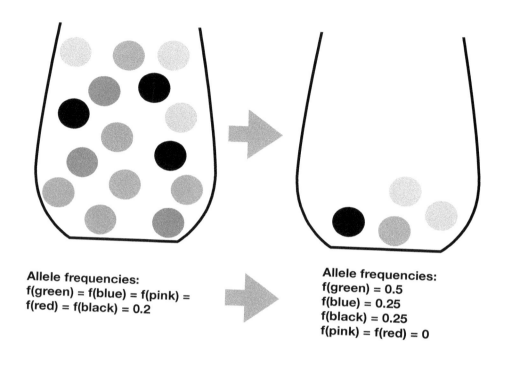

Allele frequencies:
f(green) = f(blue) = f(pink) =
f(red) = f(black) = 0.2

Allele frequencies:
f(green) = 0.5
f(blue) = 0.25
f(black) = 0.25
f(pink) = f(red) = 0

(c) A population bottleneck can sharply change the allelic frequencies in a population.

(d) The Asiatic lions found in the Gir National Park are a good example of a population bottleneck.

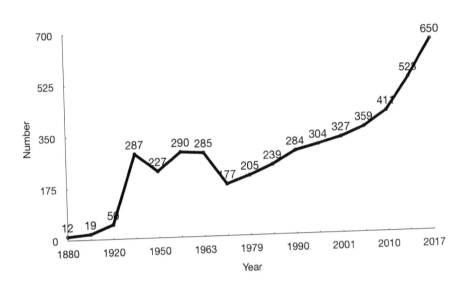

(e) The lion population shows recent population bottlenecks. While the population today is more than 600, at one time, we had only a dozen lions left. Data source: [Meena, 2009, Government, 2020].

(f) Mate preference in peacocks has led to the evolution of very large tails that serve little purpose other than display. In fact, it is much of a handicap since the peacocks loose their agility and become easy preys.

Figure 8.3: Evolution is fuelled by violations of the Hardy-Weinberg principle.

8.3 POPULATION VIABILITY ANALYSIS

Wildlife conservation needs to work at several scales. We need to preserve those individuals that are loved by people — after all, they play a huge role in generating public opinion — and funds — towards conservation. Thus, we need to work at the scales of veterinary care — sometimes even intensive veterinary care for certain high priority individuals. At the same time, we also need to work at much larger scales of preservation of habitats and habitat connectivity. For, if all the habitats are diverted or destroyed, where will the animals live? But very often, the aim of a manager of wildlife resources is to conserve populations. For instance, the Field Director of the Jim Corbett Tiger Reserve will be highly concerned to ensure the continuity of the tiger population in the Jim Corbett Tiger Reserve. For her, the survival of a particular individual tiger is probably of little importance, but the survival of the tiger population in the Jim Corbett Tiger Reserve is of immense importance — to guarantee that tigers will always exist in the Jim Corbett Tiger Reserve.

But how will she ensure the survival of the tiger population in her tiger reserve? One way is to begin by listing the threats to the tiger population. Perhaps there is a threat of poaching. Perhaps the habitat is getting degraded due to pollution. Or the habitat is getting diverted and fragmented, say for developmental activities such as setting up of dams or roads. Perhaps the tiger population has become so small that there is a risk of severe inbreeding depression, and individuals need to be brought in from outside. Or the reserve is facing severe droughts every three years or so — due to global warming and ensuing climate changes. There could be numerous challenges that the tiger population in her park faces.

Once these threats have been noted, the next step is to prioritise them. It is possible that the poaching incidents are so less that there is virtually no impact on the survival of the population, but that the droughts are so severe that unless something is done to rectify the situation (say by supplementing water and food supply), the tiger population will collapse. In such a situation, the Field Director should focus most of her attention on rectifying the drought situation.

Population viability analysis (PVA) is a procedure that helps us to make these decisions about prioritisation of activities. It analyses the viability of a population, that is, the ability of the population to persist, or to avoid extinction. In PVA, we compute the probability of the population getting extinct in a specified time — say 1,000 years. This is done by listing the threats and their probabilities of occurrence — important because the threats can occur randomly. It is possible that the threats occur one after the other, and the population has time to recover from the impacts. On the other hand, it is also possible that multiple threats strike at the same time, and the population collapses. To grasp the multitude of scenarios, a computer is often utilised to simulate the conditions — together with the probabilities of occurrence of various threats. Thus, for the next 1,000 years, multiple simulations will be made. Suppose we do 100,000 simulations, and in 15,000 of those, the population collapses within 1,000 years. In the remaining 85,000 simulations, the population survives for 1,000 years. In that case, we'll say that the there is an 85% chance of the population surviving the next 1,000 years.

Once we have the probability of survival, we can analyse the impacts of changes by comparing with the baseline survival figures. Suppose a road is being built, and it will divide the habitat into two parts. If that road gets constructed, what will be the probability of survival of the population? We do the simulations again, and here we find that out of 100,000 simulations, the population survives in only 27,000 simulations. In that case, we'll say that after the road gets constructed, there will be a 27% chance of the population surviving the next 1,000 years. Or, in other words, there will be a 73% chance that the population collapses in the next 1,000 years. That should be a major cause for concern.

But if that road *has to be* constructed, will incorporation of mitigation measures such as underpasses bring any benefit? We do the simulations again, with mobility between the two sub-populations. Suppose here we find that out of 100,000 simulations, the population survives in only

68,000 simulations. In that case, we'll say that with the road *and* underpasses, there will be a 68% chance of the population surviving the next 1,000 years. Not as good as 85% without the road, but not as bad as 27% with the road without underpasses, either.

What if we try to compensate the habitat fragmentation with incorporation of more areas in the Jim Corbett Tiger Reserve? What will be the impact of adding 30,000 hectares of land to the reserve? We do the simulations again, and here we find that out of 100,000 simulations, the population survives in only 84,500 simulations. In that case, we'll say that the addition of 30,000 hectares to compensate for the road with underpasses increases the chance of the population surviving the next 1,000 years to 84.5%, or nearly quite close to the current situation. So if the road *has to be* built, it should have underpasses *and* 30,000 hectares of land should be added to the reserve to compensate for the fragmentation of the habitat. That is the utility of doing population viability analyses — it provides *concrete* data regarding the impacts of various threats and their mitigation. Similarly we may perform a population viability analysis for different threats and understand the efficacy of various management interventions. Since the management interventions require costs, we may also use population viability analysis to do a cost-benefit analysis of the various management interventions. With this example as a backdrop, let us understand the *process* of doing population viability analysis.

8.3.1 Three ways of doing population viability analysis

The process used to perform population viability analysis depends on the amount of data available about the system (the species and its environment), the kinds of results needed as output, and the ability of the analyst. Depending on the utility, population viability analysis is also referred to as extinction risk assessment, population vulnerability analysis, predictive simulation modelling and stochastic population modelling. For instance, we may be interested in assessing the extinction probability of a single species population of wildlife [Possingham et al., 2013]. This may be done by integrating data on the life history, demography, and genetics of the species with information on the variability of the environment, diseases, stochasticity, etc. Mathematical models and computer simulations may be utilised to predict whether the population will remain viable or go extinct in a decided time frame — i.e. the probability of it getting extinct under various management options [Beissinger and McCullough, 2002]. This is an example of predictive utilisation of population viability analysis — since we are interested in predicting the fate of a population under given circumstances.

In other cases, we may be interested in knowing how to amend the situation — say by increasing the number of individuals in the population to bring it to a minimum level of viability. Thus, we may be interested in the determination of a minimum viable population, or the size at which a population has a 99% probability of persistence for 1,000 years [Shaffer, 1981]. The utility is that once this value is known, we may bring in individuals from other populations to boost the home population.

Thus population viability analysis can be used to serve different ends — predictive and manipulative. We may utilise different methods to reach the conclusions. One method is to utilise empirical observations of the stability and long-term fates of a number of populations of various sizes. An example is the study of viability of various population sizes of the bighorn sheep [Berger, 1990]. In this case, populations of various sizes were studied over time to understand whether or not — and when — do these populations become extinct. While large populations were found to survive for very long periods, smaller populations had a much greater probability of getting extinct due to small population dynamics such as inbreeding depression and stochasticity. Observations regarding the viability of different population sizes can be tabulated and extrapolated to similar species in similar environments to predict their fates.

Another method is to use analytical (mathematical) models of the extinction process to calculate the probability of extinction from measurements of parameters. A good example is Goodman's model of the demography of chance extinction [Goodman, 1987].

However, since mathematical models become intricate and quite involved with increase in the number of parameters being considered, these days we often make use of computer simulations and modelling to project the probability distribution of possible fates of a population. An example is the use of the software Vortex [Lacy, 1993]. Running simulations is dependent on the availability of two pieces of information:

1. an explicit model of the extinction process, and

2. the quantification of threats to extinction.

Of late, population viability analyses have also started to incorporate, besides the probability of extinction, other measures of the health of the population, including the mean and the variance in the population growth [Lindenmayer et al., 1995], and changes in the range, distribution, and habitat occupancy [Hanski and Gilpin, 1991] of the population, together with losses of genetic variability [Lande and Barrowclough, 1987]. These parameters help increase the accuracy of the predictions, albeit at the cost of higher computational intensity — often necessitating the use of powerful computers.

8.3.2 Utility of population viability analysis

Population viability analysis has turned out to be an important management tool for conservation [Akçakaya and Sjögren-Gulve, 2000]. Once we know how to simulate populations on a computer, we can have answers to several questions relating to

1. calculation of the probability of persistence of a population under current conditions — to make a sound judgement of the prognosis of conservation at a site,

2. calculation of the probability of persistence of a population under altered management intervention conditions — to prioritise various management interventions and to aid in doing a cost-benefit analysis of the management interventions,

3. calculation of the most likely average population conditions including population size, range of population sizes across years, and the rate of loss of genetic variation — to make decisions regarding the future course of actions for a population,

4. identification of the most likely risk factors for individual populations — such as large deaths, large fluctuations, inbreeding, or other factors — to create measures for their mitigation and/or adaptation,

5. sensitivity analysis of the population towards varying estimates of reproductive success, juvenile and adult survival, impacts of natural catastrophes, initial population size, carrying capacity of the habitat, and dispersal among populations — to better understand which factors play larger roles in the survival of the population,

6. effects of removing certain individuals from the population — say, to predict which individuals can be translocated without negative impacts to the population,

7. estimation of benefits of population supplementation through translocation or release of captive bred stocks,

8. calculation of the permissible levels of controlled harvest — to allow for planned removal of individuals,

9. estimation of the impact of poaching on a population,

10. estimation of the utility of corridors for improving the long-term viability of the population — well-functioning corridors can convert multiple small populations into a large population, helping to counter the small population dynamics in each of the individual small populations — thus the need to know the best ways to create and/or strengthen the corridor connectivity between wildlife populations, and

11. estimation of the impacts of anthropogenic activities on the long-term sustenance of populations.

Population viability analysis — still a developing field — is already playing an increasingly important role in various facets of wildlife conservation by aiding decision-making.

8.4 INBREEDING AND OUTBREEDING MANAGEMENT

8.4.1 Inbreeding

Inbreeding is the mating of closely genetically related individuals such as parents and offsprings, brothers and sisters, close cousins, grandparents and grandchildren, etc. It is a cause of concern because it increases homozygosity in the population, resulting in increased expression of recessive traits, many of which lead to genetic diseases. It also reduces variations between individuals in the population, reducing the resistance and resilience of the population to counter changes in the environment.

Inbreeding occurs naturally in certain organisms like the fruit fly (*Drosophila melanogaster*) and banded mongoose (*Mungos mungo*) where mating between close relatives is known to occur. However, they are the exceptions to the rule of avoiding incest, a rule that is observed in most species. However, with the loss and fragmentation of habitats, and reduction in population sizes, today we find increasing number of instances where animal populations are getting forced to inbreed. This occurs when the animal population is so small or isolated that most individuals are already genetically related to each other, and so inbreeding is the only option to breed — the populations that do not inbreed simply get exterminated when the older individuals die and there are no young ones to replace them!

Inbreeding — even in those species where it is naturally found such as the banded mongoose (*Mungos mungo*) — is known to cause inbreeding depression, defined as reduced survival and fertility of offsprings born of related individuals. We also find a strong correlation between genetic diseases and reduction of heterozygosity — defined as the possession of two different alleles of a particular gene or of several genes by an average individual of a species or population. The average heterozygosity of different species discerned from allelic isozyme loci [O'Brien et al., 1983] is given below:

1. *Acinonyx jubatus*: 0

2. *Felis catus*: 0.076

3. *Homo sapiens*: 0.063

4. *Mus musculus*: 0.088

We can observe that species such as the cheetah (*Acinonyx jubatus*), which have suffered multiple population bottlenecks [O'Brien et al., 1987] have negligible heterozygosity, whereas species that have not suffered large population bottlenecks, such as the domestic cat (*Felis catus*) have a considerable amount of heterozygosity in the average individual.

This lack of heterozygosity has important consequences for conservation. If we compare the seminal traits of cheetah and the domestic cat [Table 8.1], we will find that not only is the sperm density and motility less in cheetahs, but also that cheetah sperms suffer from several severe structural abnormalities. Similar features — low density, low motility and a variety of structural defects — are also found in the sperms of African and Asiatic lions [Wildt et al., 1987]. This can explain the low fertility rates in these already threatened species.

Table 8.1: Seminal traits of South African Cheetah and the domestic cat [Wildt et al., 1983].

Characteristic	Cheetah	Domestic cat
Sperm numbers per ml of ejaculate	$14.5 \pm 1.8 \times 10^6$	$147.0 \pm 39.5 \times 10^6$
Spermatozoal motility	$54.0 \pm 3.0\%$	$77.0 \pm 3.0\%$
Coiled flagellum	$25.8 \pm 2.3\%$	$5.5 \pm 0.8\%$
Microcephalic defect	$1.2 \pm 0.3\%$	$0.2 \pm 0.1\%$
Bent midpiece	$23.3 \pm 1.1\%$	$6.4 \pm 0.8\%$
Bent flagellum	$16.2 \pm 1.3\%$	$5.1 \pm 0.7\%$

Highly inbred animals such as the cheetah also suffer from a weakened immune response to diseases [O'Brien et al., 1985]. This often occurs because these animals do not have significant variation in their MHC (Major Histocompatibility Complex) genetic makeup — a set of cell surface proteins that recognises foreign molecules. With little variation in the MHC genetic makeup, the cheetahs' immune system can not recognise many pathogens as foreign bodies, and so fails to elicit a response to fight the pathogens. As a result, they easily fall sick — a fact that also explains their high mortality rates. Data of juvenile mortality in cheetahs are presented in table 8.2. We observe that cheetahs suffer from high juvenile mortality, which increases further if the population is closely related to each other — further aggravating the inbreeding depression present in this species.

Table 8.2: Juvenile mortality in cheetahs till six months of age. Data source: [O'Brien et al., 1985].

Population	Mortality (%)
Unrelated population	26.3
Related population	44.2

Inbred populations of other species — such as the Isle Royale wolves [Mlot, 2013, Mlot, 2015, Wayne et al., 1991, Räikkönen et al., 2009] — also show high levels of stillbirths (abortions) together with physical deformities such as opaque eyes and hunchbacks — making them less adept to hunting and getting food. These factors greatly reduce their fitness and the ability to survive.

There are two reasons why inbreeding depression should occur:

1. The partial dominance hypothesis postulates that inbreeding increases the possibility of individuals expressing the effects of recessive deleterious mutations [Fig. 8.4a]. While these mutations are present in low frequencies in the population, they don't express themselves when the organisms are heterozygous. Since they are present in low frequencies, we may assume that most individuals don't have these mutations, and a few are heterozygous for this mutation. In random matings, the two more likely situations are:

   (a) None of the parents has the mutation (both are *AA*). In this case, the offspring will also not have the mutation (they will be *AA*):

$$AA \times AA \rightarrow AA(100\%)$$

(b) One parent doesn't have the mutation (it is *AA*), and the other is heterozygous (it is *Aa*). In this case, the offsprings will either not have the mutation (they will be *AA*), or will not express the mutation (they will be *Aa*):

$$AA \times Aa \rightarrow AA(50\%) + Aa(50\%)$$

However, in close matings, there is an increased possibility that both parents are heterozygous (they are *Aa*). In such cases, 25% of the offsprings will express the deleterious mutation (they will be *aa*):

$$Aa \times Aa \rightarrow AA(25\%) + Aa(50\%) + aa(25\%)$$

In many cases, the effects of the deleterious mutations are in the form of severe developmental abnormalities resulting in abortions, or weak metabolism and ability to fight diseases resulting in later deaths. This explains the reduced survival of inbred offsprings. At other times, the effects of the deleterious mutations may be in the form of abnormalities in the production of sperms and ova, resulting in reduced fertility observed in inbred populations.

2. The overdominance hypothesis posits that for several genes, heterozygosity is advantageous over homozygosity. An example is the sickle cell disease caused by a recessive mutation — individuals with two mutant copies of the gene get the sickle cell disease, while individuals with two normal copies of the gene are normal but susceptible to malaria. But the individuals heterozygous for the mutation — the carriers of the disease — get an added protection against malaria — the malarial parasite is unable to reproduce in their red blood cells. This is why alleles with heterozygote advantage are maintained by balancing selection at intermediate frequencies, even when in a homozygous state they would result in a disease. Similar heterozygous advantage also occurs with several other genes. This is one reason why outbreeding creates individuals that show hybrid vigour or heterosis — improved function and quality of hybrid offsprings when compared to inbred offsprings — since many of their genes are in a heterozygous (advantageous) state.

Inbreeding increases the possibility that the offsprings are homozygous for several genes. In such conditions, they do not get the heterozygous advantage, and so have lower fitness than individuals born out of random matings.

8.4.2 Management of inbreeding

Since inbreeding occurs due to mating between closely related individuals, it can be countered by introducing genetically different individuals in the population. We have examples from nature where a single individual could rescue an inbred population of wolves [Mlot, 2013]. Scientific management of inbreeding depression is based on introducing outside individuals artificially into the inbred population(s) to achieve genetic rescue — restoration of genetic diversity and reduction of extinction risks in small, isolated, inbred populations through introduction of new alleles in the gene pool. This can be done through natural movement of animals — say by creating or strengthening their movement corridors, or by an artificial movement of individuals — by catching, transporting and releasing them into the inbred population.

There are several terms highlighting the movement of individuals from one place to another. The deliberate and mediated movement of individuals or their populations from one place to another is known as *translocation*. This may be done for *reinforcement* (also known as *supplementation*) of the population found at the translocated area, or for *reintroduction* of the species to an area

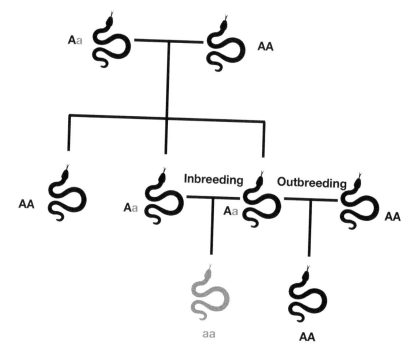

(a) An illustration of inbreeding depression in snakes. Mating between close relatives results in an individual expressing the recessive traits, some of which may reduce fitness.

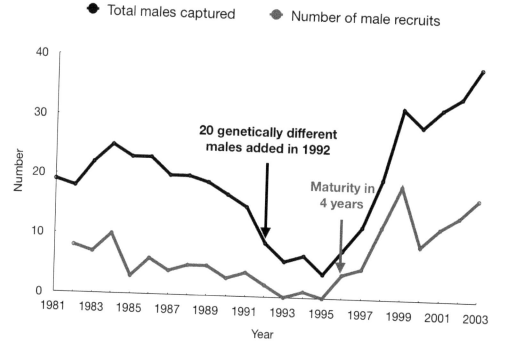

(b) Genetic rescue of an inbred European adder (*Vipera berus*) snake population.

Figure 8.4: Inbreeding depression and genetic rescue. Data source: [Madsen et al., 1999, Madsen et al., 2004].

from which it has become extinct. It may also be done for *introduction* of the species to a location where it was never found before.

In genetic rescue, we translocate a few or several individuals from one population to another population to reinforce (or supplement) the population at the target site, while also introducing new alleles to, or altering the allele frequencies of, the target population. A good example is the genetic rescue of an inbred European adder (*Vipera berus*) snake population by introducing genetically different males to the target population [Fig. 8.4b]. Before the intervention, the inbred snake population was diminishing — continuously — for several years. The number of new recruits — indicating births and the early survival of the young ones — was also going down. When individuals from another population in the form of genetically different males were added, there was a rapid growth in the population together with a large increase in the number of new recruits, demonstrating that the intervention led to a genetic rescue of the inbred population.

8.4.3 Outbreeding

When attempting genetic rescue, it is also possible to *overdo* the introduction of new alleles. Two good examples can be cited where introduction of very different individuals decimated the target population [Turček and Hickey, 1951].

The first is the story of ibex. By the 1850s, Alpine Ibex (*Capra ibex*) was extinct in the High Tatra mountain range at the Slovakia-Poland border, mostly due to over hunting. In 1901, the species was re-introduced to the areas by bringing in individuals from Austria. Then, in an attempt to increase genetic diversity, *Capra hircus* and *Capra nubiana* were also introduced. In those days, they were considered sub-species of *Capra ibex*. The introduction of these three species together resulted in the death of all the animals.

Why did they die? It was found out that the Alpine Ibex *Capra ibex* is adapted to cold climates. It ruts and breeds in the winter season and fawns in spring. Thus, their young ones are born when the harsh winter is gone and summers are approaching. On the other hand, *Capra hircus* and *Capra nubiana* are adapted to warmer climates. They rut and breed in the autumn season and fawn in winter. Thus, their young ones are born when the harsh summer is gone and it is getting colder. When the animals inter-bred, the *Capra ibex X hircus X nubiana* individuals got their breeding traits from their warm-climate-adapted parents. Thus, they rutted and bred in the autumn season and fawned in winter. Their offsprings perished from generation to generation in the severe winters.

The second story is that of the roebuck. Before World War I, the Siberian race of roebuck *Capreolus capreolus pygargus* was introduced in Slovakia. The Siberian race is much larger than the Slovakian race. When males of the Siberian race mated with females of the Slovakian race, the foetus was so large that fawning became impossible, and the females died.

Thus, outbreeding — the practice of introducing unrelated genetic material into a breeding line — can result in tragic consequences. This is referred to as outbreeding depression — the decreased fitness in hybrid offsprings when compared to their parents, or when compared to offsprings from crosses between more related individuals. Outbreeding depression often manifests in two ways:

1. such intermediate genotypes are formed as are not adapted to either parental habitat, such as in the case of ibex, or

2. there is a breakdown of biochemical or physiological compatibility in the hybrids, such as in the case of roebuck.

Thus, management of outbreeding also becomes important.

8.4.4 Management of outbreeding

When bringing in individuals from an outside population, several things can go wrong. Thus proper management becomes important [IUCN, 1998]. It is often vital to act with speed — say to prevent capture myopathy in herbivorous animals [See section 7.6]. Thus, approvals from government agencies, landowners, etc. as required must be in place, and sufficient funding arranged for suitable backup equipment. Transport plans must be drawn up and determination of release strategy — hard or soft release — made beforehand. Education, training, and involvement of locals, together with establishment of intervention policies must also be decided and executed before the operation. Since animals from outside population are likely to bring their diseases with them, their health and genetic screening, together with required vaccinations, will have to be done, and quarantine arrangements set up.

It is advisable to do a 'test' breeding with few individuals before attempting a large-scale population mix-up. This permits analysing outbreeding depression and other hazards. Thus, the animals must be moved in several stages — the first being that of a small sample for test breeding. The released individuals need to be monitored through tagging, telemetry, informants, and other methods. Demographic, ecological, and behavioural studies of the released stock will give indications of how the introduced individuals are behaving, adapting and breeding with the local population. If any mortalities occur, they must be swiftly and thoroughly investigated to discern the possibilities of diseases and outbreeding depression. At times, interventions such as supplemental feeding and veterinary aid may be required.

When we are working with threatened species, every individual counts. Thus the managers of the outbreeding activity must have the power to revise and reverse decisions, reschedule, or even discontinue the programme, should the need arise. Since many threatened species are those that were earlier wiped out through hunting — say because they were ferocious, harmed crops, or were considered 'unclean' or otherwise culturally unwanted — extensive public relations activities, including education, outreach, and mass media coverage may also be needed.

Obviously, population supplementation will not work if suitable habitats are lacking. Since the aim of outbreeding is often to increase the fertility and viability of the population, we need to make arrangements for the housing of the enlarged populations. Thus, habitat protection and restoration activities [see Chapter 9], including strengthening of historical corridors, become crucial. The evaluation of success or failure of the program, together with the cost-effectiveness of the techniques used must be made in an unbiased manner, and published, to guide similar breeding management operations in the future.

Habitat management

We observed in Chapter 4 that habitat is the "subset of physical and biotic environmental factors that permit an animal (or plant) to survive and reproduce [Block and Brennan, 1993]." We have also discussed how the degradation, loss, and fragmentation of habitats are putting huge pressures on the survival of numerous species — without their habitat, the species cannot exist. The need of the hour is to save and restore the habitats that remain, by closely monitoring them for any degradations and taking quick and adequate remedial actions in the form of habitat management. In this chapter, we discuss these topics in detail.

9.1 HABITAT COMPOSITION AND IMPORTANCE

A habitat is comprised of biotic (living) and abiotic (non-living) components. These components together provide resources and services to the species living in the habitat. They provide things such as food and water, space for living, and cover — or shelter — from weather and predators. The cover may be a protective cover, such as a breeding cover, an escape cover, and a thermal cover, or may be a hunting-related cover such as an ambush cover. While both prey and predator species require protective covers, ambush covers are only used by the predators when they hunt by stealth.

By constitution, the covers may be vegetal covers — provided by plants such as trees, shrubs, grasses, etc. Or they may be non-vegetal covers — provided by rocks, caves, cliffs, burrows, etc. The vegetal covers are part of the vegetal component of the habitat [Fig. 9.1] — including not only the plants such as trees, shrubs, and herbs, but also snags (dead and dying trees, often with holes and crevices that are used for perching and roosting) and downwood (decaying branches and other wood on the ground). The non-vegetal covers are part of the non-vegetal component of the habitat [Fig. 9.2], which includes air, water, soil, rocks, cliffs, caves, space, and human influences.

The importance of habitats is two-fold: they sustain wildlife, and they are beneficial to humans. Taken together, the vegetal and non-vegetal components of the habitat provide all that is required for the inhabiting organisms to survive and sustain. At the same time, well-functioning habitats and their components also provide several benefits to humans. For example, a wetland habitat — just by virtue of its existence — acts as a reservoir of water which may also be used by humans — for drinking, irrigation, industrial, or domestic purposes. The organisms living in the wetland — both plants and animals — act on various pollutants and wastes that arrive to, or form in, the water body, thus keeping it clean. Wetlands also replenish the groundwater and control floods by providing a storage for excess water in the rainy season. They play a major role in regulating and tempering the local climate.

Habitats also provide food — to humans [Fig. 9.3a] and their livestock. They are our sources of numerous wood and non-wood forest produce such as fibres, fruits, peat, and medicinal products. The biodiversity in well-functioning habitats puts a check on pests and diseases through biologi-

(a) Trees in Timli Forest Range, Uttarakhand.

(b) Shrubs in Sariska Tiger Reserve.

(c) Herbs in Kruger National Park.

(d) Snags for perching in Kruger National Park.

(e) Snags for roosting at FRI Dehradun.

(f) Downwood in Timli Forest Range, Uttarakhand.

(g) Downwood in Kruger National Park.

Figure 9.1: Some examples of vegetal components of habitats.

(a) Space allows for existence, movement and activity.

(b) Special features like cliffs observed in Panna Tiger Reserve.

(c) Disturbances caused by anthropogenic activities, as observed in Kruger National Park.

Figure 9.2: Some examples of non-vegetal components of habitats.

cal control mechanisms. These are the same mechanisms that naturally regulate the population of several species — ensuring that the carrying capacity is never exceeded.

Habitats such as the forests conserve soil, stabilise banks and act as carbon sequestration and storage sites. Good habitats are areas of nature studies, recreation, tourism [Fig. 9.3b,c], worship and beauty. Many habitats have spiritual and religious values [Fig. 9.3d] and are of contemporary cultural significance. Their role in supporting and enriching the lives of people can never be overstated.

But the use of habitats by humans has also led to the *overuse* of habitats by humans, leading to the problems of habitat degradation, loss and fragmentation. This makes it crucial to monitor and manage habitats — to conserve them and to keep them in a well-functional state.

9.2   HABITAT MONITORING

Monitoring may be defined as "the collection and analysis of repeated observations or measurements to evaluate changes in condition and progress toward meeting a resource or management objective [Rowland and Vojta, 2013]." It includes making an assessment of the information needed, planning and scheduling the collection of data to derive the desired information, followed by the actual collection of data. The data is then classified, stored, mapped or otherwise evaluated to generate information, which gets reported.

On the basis of purpose, monitoring can be of several kinds:

1. Targeted monitoring — It is used when the 'targets' are known. Often the targets are priority species, priority habitats or certain characteristics of the habitats (such as salinity or oxygen concentration). Once the targets are well defined, target monitoring keeps track of the condition of the targets, especially in the context of their response to management interventions.

2. Cause-and-effect monitoring — It is used to analyse the impacts (effects) of changes (causes) emanating from management interventions or disturbance to the habitats. It is often used when a change in the *status quo* of a habitat is imminent — say when a dam is being constructed or an area is being mined; the cause-and-effect monitoring will investigate the effects of these activities.

3. Context monitoring — It addresses a wide array of ecosystem components at multiple scales. There is no specific reference to influences of ongoing management.

There are five steps to designing a monitoring program. They are:

1. Development of the objective of monitoring: We need to objectively delineate why we are doing the monitoring exercise. What kinds of changes do we anticipate and wish to analyse? This step helps us decide the data parameters — area, species, characteristics, etc. that will best recognise those changes in a timely and cost-effective manner. It also aids in deciding the time intervals in which data should be collected to recognise the anticipated changes.

2. Assessment of existing data — literature review: This is required because some of the data may already be available from other sources, fulfilling the objective of monitoring. In such cases, these data may directly be used (if they still suit the objective of monitoring), and time and resources utilised only for gathering those data that are unavailable.

3. Planning a design for the collection of new data (which may be field data or remotely-sensed data): We need to decide whether we need a census or a sample. If a sample, what should be the sampling intensity? Should we go for a random sampling, a systematic sampling, a stratified sampling or a multi-stage sampling? How are we going to collect the data — in

(a) Extraction of resources such as fishes forms the livelihood of several people living near the wetlands. This image shows locals fishing in a lake in Alwar district of Rajasthan.

(b) Sea beaches, such as the Radhanagar beach in Andaman Islands, are a major tourist attraction.

(c) The Nalsarovar Bird Sanctuary is a photoshoot destination.

(d) Dev Prayag in Uttarakhand has religious and cultural significance as the point where Alaknanda and Bhagirathi rivers join to form the Ganges.

Figure 9.3: Examples of benefits from well-functioning habitats.

the form of pen-and-paper forms, photographs, or laboratory samples? How are we going to process the data? How often will the data be collected? How do we plan the logistics? A good plan will help us answer these and several other questions before we embark on the actual collection of data.

4. Running a pilot study: Once we have the plan, it is now time to do a pilot study on a small area, just to check if the plan works. When we do the pilot study, we may come to know that an extra parameter may be required in the forms, or that it is impossible to note certain characteristics just from photographs — we also need field measurements. In this way, the pilot study helps us understand the lacunae in the planned design and provides us an opportunity to rectify the lacunae.

5. Incorporating the design decisions in the monitoring program: Once the pilot study is done and changes have been made in the design plan, we can incorporate the final design into the monitoring program that will be implemented.

A monitoring program often has six components:

1. Introduction: This section explains the objectives of monitoring and the anticipated changes in the system being monitored. Thus, it helps to bring the reader to the same page as the person who has designed the monitoring program.

2. Planning and design: This section explains the design of the monitoring program and the plan that will be used to monitor the system. It may include details about the various alternatives that were explored during the planning process, and the results of the pilot study. In this way, it removes doubts and uncertainties regarding the alternatives.

3. Data collection: This section describes how the data will be collected — the equipments required, the characteristics that will be measured, how they will be measured, and so on. This section often provides model formats of data collection so that every person following the monitoring program collects the data in the same format — aiding later analyses.

4. Data storage: This section describes how the data will be stored. Often the collected data are digitised or entered into digital platforms for aiding analyses. Will this be done in the current monitoring program? If yes, who will do the data entry — the person who collected the information, or another person? And when will the data be entered — at the same time the data is collected (say using a tablet computer), at the end of the day, or at the end of each monitoring period (which may be a few days to a few months)? These questions become important because people may misread others' handwriting or shorthands. Similarly, procrastination may lead to a loss of data. Once the data has been digitised, how will it be stored? What are the data backup plans? And what happens to the original data sheets — are they stored for a later cross-checking, or destroyed after digitisation? This section answers these questions, together with the logic used to arrive at the answers.

5. Data analysis: This section details the usage of the data. It answers questions such as: Once the data is available, how will it be used to discern information? What kinds of analyses will be done? Which algorithms will be followed? Will the analyses be done manually or will they be automated? Who will do the analyses — the collector of the data or a central agency?

6. Reporting schedule (yearly, 5-yearly, 10-yearly, etc.): This section details the schedule of reporting, i.e. how often will the data collection and analyses be done. For systems that are anticipated to change quickly, a monthly or yearly data-reporting schedule is preferred. An

example is the changes to the habitat brought about due to setting up of a new mine — because of pollution, we may expect a rapid degradation of the habitat. For systems that are anticipated to change slowly, a 5-yearly or decadal reporting schedule may be preferred. An example is monitoring the changes in a habitat due to global warming.

A habitat-monitoring program collects data on several attributes. For example, in a forest area, it may monitor the vegetation composition, including vegetation type (% of area under each vegetation type, e.g. forest, grassland, etc.) and species abundance (in terms of number of individuals of each species, frequency of occurrence, biomass, etc.). It may also include data on vegetation structure of trees (including canopy cover, canopy closure, tree diameter, basal area, tree height, stand density, etc.), snags (including decay class, diameter, height, density (number per unit area), size of cavities, etc.), downwood (including decay class, diameter, length, cover provided, density (number per unit area), volume, etc.), shrubs (including shrub cover and shrub height), herbs (including herb cover and herb height), and seral stages.

In the case of a wetland such as the Chilika lake [Fig. 9.4a], the monitoring program incorporates collection of water samples from different locations and depths [Fig. 9.4b] that are then analysed in a laboratory [Fig. 9.4b] for water composition, salinity, organisms [Fig. 9.4c], etc. The birds in the wetland and surrounding areas are often captured and tagged for tracking purposes [Fig. 9.4d,e], and detailed records are maintained and analysed.

In this context, we must also appreciate the challenges to monitoring. The delineation of all the relevant attributes defining the habitat of a species is difficult, and so most monitoring programs capture only a subset of the relevant attributes. With an increase in our understanding of various habitats and technological innovations permitting vast and automated data capturing, the number of attributes captured in habitat monitoring programs are on the rise, and we may expect them to further increase and improve in the future. Similarly, deficiencies may remain in deciding and implementing the correct spatio-temporal scales for monitoring — and consequently the collected data may be insufficient or superfluous for discerning intelligible information. Taking accurate measurements is difficult, especially in field situations, and we must take all results with an understanding of the errors that may have crept in the measurements and during data processing. Often the cause-effect relations are incompletely understood, and it is possible that wrong causations get implied. Similarly, discerning the relative roles of the habitat and other factors (e.g. competition, life history traits, etc.) in the response of populations may be difficult to delineate. In such cases, if we used the population response as a proxy for the measurement of habitat characteristics, we may arrive at incorrect conclusions.

Thus, while a robust monitoring protocol is *sine qua non* for habitat management, we must recognise that no monitoring is perfect — there is always a scope for improvement.

9.3 HABITAT MANAGEMENT AND IMPROVEMENT

Suppose, through habitat monitoring procedures, we realise that the habitat is good. In such a case, the habitat must be *maintained* in the good state. Only routine housekeeping operations — such as protection and slight manipulations here and there — may be needed. On the other hand, we may come to the realisation that the habitat is rapidly getting degraded, or has already become extremely degraded. In such a scenario, the habitat must be *improved* — brought back to a good state through manipulations and active management. What are the options available to us to manage a habitat? And how can we improve a habitat? To manage and improve habitats, we must understand and make use of the *tools of habitat management*.

(a) The Chilika lagoon is characterised by shallow brackish waters.

(b) The habitat of Chilika is continuously monitored by taking samples from different locations and depths.

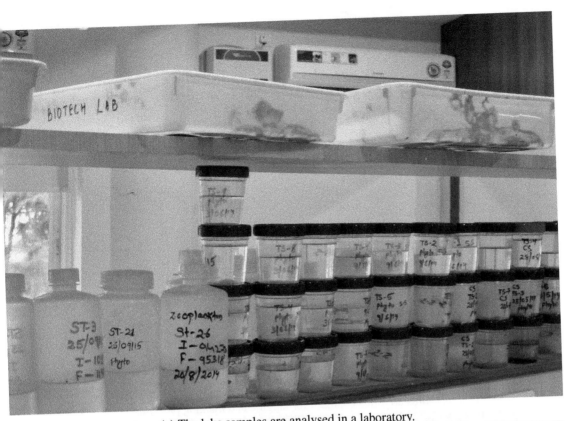

(c) The lake samples are analysed in a laboratory.

(d) The monitoring of biological samples is done after staining.

(e) Birds are often captured for tagging and tracking purposes.

(f) Plastic and metal monitoring tags are attached to the legs of birds.

Figure 9.4: Habitat management at Chilika lake.

9.3.1 Tools of habitat management

Early thinking about habitat management on scientific lines can be traced to Aldo Leopold, who, in his classic work [Leopold, 1933] formulated five tools for the management of habitats [Fig. 9.5]:

1. Axe — used to remove those vegetation that are not needed — such as invasive species

2. Plough — used to work soil for raising plants — especially native fodder species

3. Cow — used as a means of removing grass

4. Fire — used as a means of restarting succession

5. Gun — used to remove those animals that are not needed — such as invasive species or sick individuals in overpopulated stocks

Figure 9.5: Aldo Leopold's five tools of habitat management.

It is easy to see that Aldo Leopold's tools of habitat management are also the *tools of habitat destruction*. The key is to understand *how to use them*.

Let us consider the example of livestock (cow). When unmanaged, livestock bring several negative impacts to the habitats. They compete with wildlife for space, fodder, water, etc., often leading to displacement of wildlife to non-prime/sub-prime habitats such as hills or rocky patches. Over grazing leads to degradation of the habitat, sometimes even complete loss of the habitat. Most of the wildlife corridors and migratory routes have been encroached upon to raise cattle — leading to severe fragmentation of wildlife habitats. Cattle change the behaviour of wildlife — especially by causing disturbance to mating and fawning activities. They spread diseases like ringworm, tapeworm, foot and mouth disease, and tuberculosis, trample the nests of ground-dwelling birds and reduce the nesting sites available to them, compact the soil through their movement, and pollute the water bodies often leading to eutrophication. Also, when wild animals attack livestock, they often lead to situations of conflict — and people kill the wild animals in retaliation. Thus, cattle truly are a menace to wildlife management.

But this is so only when they themselves are *unmanaged*. If we use them properly, cattle can do wonders. Cattle are often indiscriminate feeders, and so can be used to remove coarse tall grasses — allowing soft palatable grasses to grow, thus improving the forage quality. We've seen in Chapter 2 that in the process of grazing, cattle make insects available to birds such as egrets. When a grassland is getting transformed into a woodland through ecological succession, regulated grazing can be used to remove plants of the tree species and thus maintain the area as a grassland for wildlife. The removal and reduction of canopy cover often benefits small rodents, as well as birds of prey. Selective, patchy grazing can be used to create structurally dense habitats with lots of ecotones and species diversity, and to remove weed plants. We can also use selective grazing to create travel corridors for animals, or to reduce the fuel load in a forest, and thus the fire hazard. And we can do all these, and more, without the application of chemicals or machines — just by using livestock *properly*.

Thus, the tools are not bad themselves, but it is the manner in which they are used. Use them indiscriminately, and the habitat gets destroyed. Use them judicially, and they become an asset to habitat management. Good habitat management always strives to reduce the negative consequences of these tools, while promoting the positive ones.

9.3.2 Activities taken up for habitat management

When the habitat is good, it needs to be maintained and protected. One activity commonly required is the control of unregulated fires. Forest fires are an integral part of forest ecosystems — especially of deciduous forests. However, unregulated fires can wipe off large chunks of habitats and lead to heavy mortality of wildlife. Hence, we need to *manage* the forest fires — and ensure that they do not spread in an unregulated manner — at wrong times and places. For this, we routinely make use of satellites and devices installed in the forests to get near-real time information about forest fires. Data about forest fires are collated to form fire maps [Fig. 9.6a] that indicate which areas are at a greater risk of forest fires, and which areas are at a lower risk. Such maps may also be utilised in forest fire forecasting where computers are utilised to predict forest fire locations depending on the current temperatures, fuel load, fire history, and wind speeds. The areas with a high risk of unregulated forest fires are often treated by construction of fire lines [Fig. 9.6b] and fire breakers [Fig. 9.6c] to divide the forest into chunks, ensuring that a fire at one location in the forest does not spread and consume the complete forest. Regulated burning of vegetal matter — especially in cold and damp seasons — may also be used to reduce the fuel load in high risk areas.

Another activity is to control the spread of invasive species, especially the non-native, alien species. We often refrain from using chemical weedicides since they may have unintended and long-term consequences on the ecology. The invasive species are generally removed manually [Fig. 9.6d] or by using machines [Fig. 9.6e]. The freed areas get quickly reoccupied by the native wildlife [Fig. 9.6f].

Grassland management may require uprooting of trees that are coming up through ecological succession [Fig. 9.6g]. While conversion of grassland into woodland is a natural process, the maintenance of grasslands as grasslands may be needed to support grazing species. This is commonly done in many tiger reserves.

During times of extreme droughts, provisioning of waterholes [Fig. 9.6h] may be needed to provide water to wildlife. This is becoming especially important due to changes in the climate, which are making several regions drier than before. In areas where a heavy leaching of salts has occurred, or where salts have been mined and removed by humans, artificial salt licks [Fig. 9.6i] may have to be made. These artificial salt licks also play important roles in the management and control of wildlife diseases, for essential minerals, and sometimes medicines, may be provided to animals by mixing them in the salt.

Good wildlife management also requires involvement of locals and stakeholders. They can provide useful information about encroachers, poachers and other wildlife offenders — sometimes even before the offence gets committed. At times, their help may also be needed to combat forest fires and remove invasive species. Thus, wildlife management integrally involves interacting with the locals and various stakeholders to create a dialogue and a win-win situation for them and the wildlife. Often this requires regular communication, shared activities, goodwill measures such as health camps [Fig. 9.6j], and timely and adequate compensation for any damage caused to them by the wildlife. Information from the locals is often augmented by the use of technology. Devices such as camera traps and autocams [Fig. 9.6k], and satellite imagery [Fig. 9.6l] can give us near-real-time information about illicit entry of poachers and encroachers.

The habitats are often assisted through plantation drives [Fig. 9.6m] and collection and disposal of trash and rubbish [Fig. 9.6n]. Priority animals are also given adequate veterinary care [Fig. 9.6o] to prevent the spread of diseases in their population(s). They may also be provided supporting infrastructure on a case-by-case basis [Fig. 9.6p].

Most importantly, the habitats need to be regularly protected by deploying protection personnel [Fig. 9.6q] and setting up of patrolling and protection camps [Fig. 9.6r] in crucial areas.

9.3.3 Options for improving habitat

At times, the habitat is extremely degraded and needs to be improved. Good examples are forests after a severe forest fire event, and areas received after completion of mining — which are often not just devoid of vegetation, but also have large pits and high levels of soil toxicity. In these cases, the 'habitat' needs extensive 'working' to bring it back to a habitable state. Different options of habitat improvement may be employed depending on the condition of the habitat [Fig. 9.7]:

1. Recovery/Neglect: This is the simplest of all habitat improvement options, and involves letting nature take its own course. With the passage of time, the degraded habitats will be colonised and the process of ecological succession will begin. A good example is leaving a harvested forest as it is so that the plants re-occupy the area.

   The benefit of this approach is that it does not require funding and active working in the habitat. Also, it permits the species best suited to the current conditions to occupy the area. However, it often requires lots of time to improve the habitat, and also suffers from the risk that in place of improving, the habitat may actually degrade further when it is left neglected. A good example is the loss of soil cover due to erosion when the land is left fallow. Without a cover of soil, the habitat may reach a much more degraded state than it was in before.

2. Restoration: This approach tries actively to return the habitat to its original state. Thus, in the case of a mixed forest that has been burnt in a forest fire, restoration will try to bring the forest back to the original state through active working (soil and moisture conservation, planting, etc.). Those species will be planted, and in those proportions, as existed in the original forest before it suffered degradation.

3. Rehabilitation/Reclamation: This approach aims to shift the degraded habitat towards one with a greater value, which may not necessarily be the original state of the habitat. Thus, in the case of a forest destroyed by a forest fire, this approach will try reforest the land, since a reforested land has greater value than bare land. However, it will not strive to bring the forest to the original state. The species that are available and suit the site conditions will be planted. These species may even be exotic species that show a fast rate of growth — to cover the bare land as soon as possible.

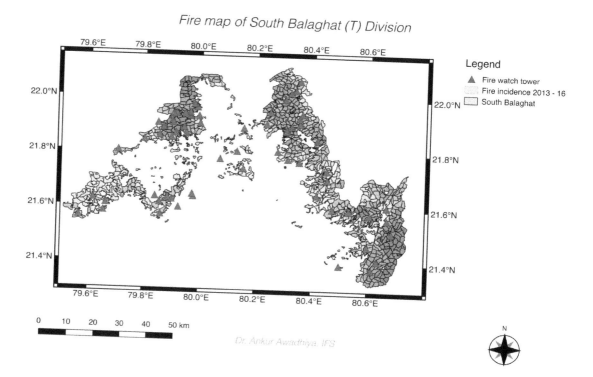

Fire map of South Balaghat (T) Division

Legend
▲ Fire watch tower
Fire incidence 2013 - 16
South Balaghat

Dr. Ankur Awadhiya, IFS

(a) Fire maps aid in rational localisation and deployment of fire-fighting personnel and equipment.

(b) Fire lines, such as these observed in Kanha Tiger Reserve, help prevent large spread of forest fire.

(c) Fire breakers are simple mud-based structures that can greatly increase the efficacy of fire lines.

(d) Removal of invasive species in Laugur Forest Range, Balaghat.

(e) Invasive alien species such as *Lantana camara* are often removed in large quantities in Mudumalai Tiger Reserve.

(f) The freed areas are quickly reclaimed by native wildlife.

(g) Uprooting of trees and controlled burning of grasslands are important parts of habitat management at Kanha Tiger Reserve.

(h) Artificial water holes, such as this one seen in Mudumalai Tiger Reserve, aid the animals in drier seasons.

(i) Artificial salt licks, such as this one seen in Buxa Tiger Reserve, are used to attract animals for treatment, and to overcome their nutritional deficiencies.

(j) Involvement of locals through health camps creates and nurtures stakeholders for wildlife conservation.

(k) Continuous habitat monitoring, through use of devices like trap cameras or AutoCams, is an integral part of habitat management.

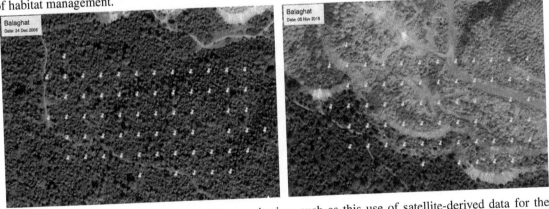

(l) Remote sensing may be used for habitat monitoring, such as this use of satellite-derived data for the identification of deforestation hotspots.

(m) Plantation drives in corridor areas promotes wildlife connectivity.

(n) Trash collection helps reduce habitat degradation.

(o) Provisioning of veterinary care, especially to priority species, forms a component of habitat management.

(p) Priority animals are provided specific support on a case-by-case basis. Here we observe artificial nests for penguins at 'Boulders' Table Mountain National Park.

(q) Adequate protection of wildlife habitats is a must for habitat management.

(r) Adequate protection of wildlife habitats is a must for habitat management.

Figure 9.6: Some examples of activities for habitat management.

Rehabilitation/reclamation is preferred over restoration when restoration is either very difficult or will require a very large amount of time and resources. During the time it takes to arrange the resources, it is possible that the area will degrade even further, say due to soil erosion or occupancy by invasive alien species. Rehabilitation/reclamation may even be done with a final objective of restoring the original habitat at a later suitable time when more resources become available.

4. Enhancement: This approach aims to improve the value of the habitat for wildlife, say through the construction of water holes for animals. This approach is often used when the habitat is not too degraded, and/or tied funding is available for these specific activities.

5. Replacement: This approach tries to create a new habitat in place of the degraded habitat. A good example is the replacement of a forest area with a marshy wetland after a mining operation is completed:

e.g. Forest $\xrightarrow{\text{Mining}}$ Mine pit $\xrightarrow[\text{Water filling}]{\text{Earth work}}$ Marshy wetland

Such an approach may be fruitful when the number and size of deep pit-holes is so large that it is economically unviable to fill them back with earth. When a marshy wetland gets created, it will be used by certain species and start playing a hydrological role, a prospect much better than barren polluting pit-holes that are a threat to wildlife and the environment.

While replacement may be more cost-effective than restoration and rehabilitation, we must ensure that this does not create a new threat to wildlife. Thus, before the pit-holes are filled up with water, they must be treated — say with a layer of soil — to ensure that wildlife is not exposed to toxic minerals and these wastes do not get leached into the groundwater, or reach another water body. Thus, any replacement should be attempted with considerable care, especially because it tries to create an exotic habitat through unnatural means.

The options of habitat improvement have become especially important as *mitigation options* for proposed developmental activities. Often when developmental activities such as construction of roads, dams and mines cannot be avoided — and large destruction of habitats is foreseeable — protection, restoration, rehabilitation and management of other habitats in lieu of the one being lost may be proposed, often in ratios $\gg 1$. In this way, the total habitat available for wildlife remains constant, or even increases. Similarly, when restoration of habitat is difficult due to permanent nature of the work, replacement of another nearby degraded site may be done in lieu. When this is done, the available options must be carefully evaluated to ensure that the best interests of wildlife (and thus, of people) are protected, and the proposed activities actually benefit the habitats — and not harm them in the zeal and guise of 'active management.'

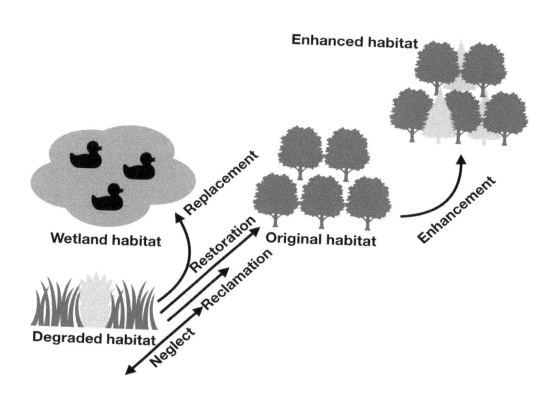

Figure 9.7: The options available to improve the existing habitat.

CHAPTER 10

Ex-situ and in-situ conservation

We began this book by noting that we need to conserve wildlife because they provide us with certain benefits [Ref: Chapter 1]. But we also observed [Ref: Chapter 4] that the species on this planet are currently facing a large number of threats, endangering their survival. We estimated in Section 4.2 the rate of loss of species in the world, and the estimate is quite humongous.

When species become extinct, the benefits that we humans get from them also end. In many cases, the resultant ecological imbalances have long-term consequences — largely beyond our capacity to adjust and cope. A good example is soil erosion in the hills. When the forests on a hill are destroyed, the soil gets exposed and begins to erode at a rapid rate. Once the soil gets removed, the land becomes bare. There is no economically viable way in which we can restore the soil that is lost. After all, lifting huge masses of soil up the hill is prohibitively expensive! Thus, the only option with us is to *conserve* the soil, by conserving the forests. And conserving the forests requires us to maintain the ecology in balance, with abundant biodiversity playing different roles. After all, we won't have sustained forests without pollinating insects and birds, without dispersing birds and monkeys, and without predators like tigers and wolves keeping a check on herbivore populations. When any species becomes extinct, the ecological resistance and resilience get chipped away, slowly but steadily. Beyond a level, the system will have such little resistance that even small changes in the environment will become threatening. And in the paucity of resilience, the system will never be able to recuperate. Thus, to reap ecological benefits, we need to conserve ecosystems. And to conserve ecosystems, we need to conserve plants and animals — often in large numbers and diversity.

But conservation requires resources — time, money, manpower, etc. It requires green decisions — which might necessitate making tough choices, such as foregoing electricity from a proposed hydropower plant, or foregoing jobs and money from a proposed mine in the hills. The dearth of resources for conservation, and the difficulty — often reluctance out of greed — in making green choices means that pragmatically we need to *economise* conservation — take such steps that conserve large areas and numerous species with minimal resources. We also need to *prioritise* the species and the areas that we are going to conserve — because of the sad reality that not all species and areas can be conserved with the minimal resources at our disposal. In this chapter, we discuss the process of making these choices and implementing them.

10.1 PRIORITISING SPECIES TO CONSERVE

"With the scarce resources available for conservation, which species should be granted protection?" If this question is put to different experts, they may have different views. A tiger researcher might say that the tiger is the most important species — the lion's share of resources should be used for the conservation of tigers. On the other hand, an ornithologist might argue that we've already spent so much resource on tiger conservation while neglecting other species — perhaps now is the time for the birds! An entomologist might plead the case for insects — for they are the biggest pollinators and they sustain the forests. All of these are perfectly logical arguments from the perspective of individual domain experts.

So how do we choose? We cannot leave the choice of species to be conserved to one — or few domain experts — for they'll come up with extremely divergent views. But there has to be a logical choice — and soon. This brings us to the concept of keystone, umbrella and flagship species.

Keystone species are those that play *critical* ecological roles — often those that are of much greater importance than we would predict from their numerical abundance [Power et al., 1996]. A good example is the banyan tree (*Ficus benghalensis*). A single banyan tree — because of its vast branches — can provide residence to a large number of birds, mammals, reptiles, and insects. Nearly all parts of the tree are edible — from fruits and flowers to leaves and branches. Thus, it is a perennial source of food to a vast variety of organisms. It even provides food when other sources of food go dry — say in the peak summer season. Thus if we make a plot of the biodiversity in a forest, the single banyan tree will light up as a storehouse of immense biodiversity. In this way, its importance to the forest is much greater than its numerical abundance — even one tree is very important. Thus, keystone species *must* be afforded protection — since by protecting them, we can protect and support a very large biodiversity. Another example is mangroves — that support numerous species of fishes, crabs, and mammals.

Keystone species also include ecosystem engineers — such as the beavers (*Castor canadensis*) and the elephants (*Elephas maximus*) [Fig. 10.1a]. Beavers construct dams — converting streams (lotic habitats) into ponds (lentic habitats) — thus creating numerous ecotones that support a large biodiversity. Elephants uproot trees — and in the process maintain grasslands by preventing their conversion into woodlands. And being destructive feeders, they support other herbivores by bringing to them the leaves and fruits of the trees that have been uprooted — a very crucial role during seasons of food scarcity in the forest.

Predators such as wolves (*Canis lupus*), tigers (*Panthera tigris*) [Fig. 10.1b], and lions (*Panthera leo*) [Fig. 10.1c] play important roles in regulating herbivore numbers and density, permitting young plants to thrive — and thus sustaining the forest. They, too, are keystone species.

Umbrella species are those species that have very large home ranges — often encompassing multiple habitats. A good example is the elephant whose home range is several hundreds of square kilometres. In the case of Siberian tigers, the home range may reach a few thousand square kilometres. When we work to conserve such species by protecting their habitat, all the species that live in their large home ranges will also get protected — automatically, without the need for any extra effort or monitoring! This makes them economically efficient target species to be conserved — facilitating a large amount of conservation with very little monitoring costs. So they must also be a conservation priority, for they bring the maximum "bang for the buck" — extremely consequential since we are always short of resources for conservation.

Flagship species are those that bring money and people for conservation. They are well known charismatic species that have captured the public's hearts and minds. They can be cute species such as the giant panda (*Ailuropoda melanoleuca*), or awe-inspiring species such as the tiger and gorilla (*Gorilla gorilla*), or species with religious and cultural overtones — such as the elephant. We often find these animals represented in the logos of several companies and organisations — precisely because they help them connect to people. Conservation of such species wins lots of public support, and brings funds for conservation. They *must* also be a conservation priority, for it would be a real disgrace if they turn extinct.

To sum up, we need to conserve the keystone species, the umbrella species and the flagship species. But what if we needed to choose between them? In that case, the winners will be those species that have many of these traits — or all of these traits. Thus, most of the attention for

(a) Elephant (*Elephas maximus*).

(b) Tiger (*Panthera tigris*).

(c) Lion (*Panthera leo*).

Figure 10.1: Examples of species that are keystone, umbrella, and flagship species.

conservation is given to those species that meet all three definitions. They are important for their habitats, their protection automatically protects several other species as well, and they are able to generate enough funding and support for the cause of conservation. Examples include the tiger, the elephant and the lion [Fig. 10.1].

10.2 CONSERVATION OPTIONS

Once we have selected the species to conserve, how do we actually go about conserving them?

In this regard, there are two main routes to conservation — *ex-situ* conservation, meaning conservation *off-site*, or outside the natural habitat of the animal, and *in-situ* conservation, meaning conservation *on-site*, within the natural habitat of the animal. *Ex-situ* conservation can take such forms as zoos, aquaria, captive breeding facilities, botanical gardens, bambuseta, arboreta, seed banks, and cryopreservation facilities — tissue cultures, sperm banks, ova banks, etc. *In-situ* conservation generally involves creation of wildlife preservation areas — wildlife reserves — where the animals are afforded protection in their natural habitat(s).

Which of these will be chosen depends a lot upon the species — they often respond differently [Fig. 10.2a]. Thus, while tigers can be conserved through any of the two routes, elephants only do well in *in-situ* conservation. In captivity, their populations slowly go down.

The choice will also depend upon the available resources. *Ex-situ* conservation is typically more expensive than *in-situ* conservation [Fig. 10.2b], especially for large-bodied animals that require more resources [Fig. 10.2c].

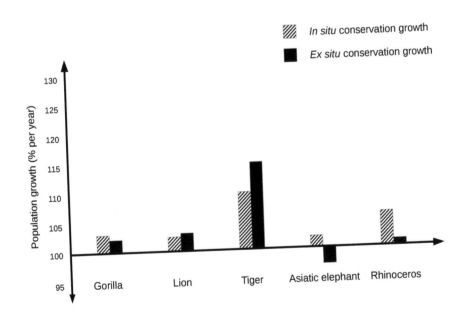

(a) Growth rates of different priority species.

There are also other considerations to keep in mind. *Ex-situ* conservation areas can provide more urgent and more intensive interventions — typically required for critically endangered species whose habitats are already largely lost. *Ex-situ* conservation may be the only option left for these species. Similarly, there are species that show intricate behaviours — such as tool usage by chimpanzees. These behaviours often get lost when the animals are raised in captivity. For such species, *in-situ* conservation may be preferable.

With this background, we now examine both the conservation routes.

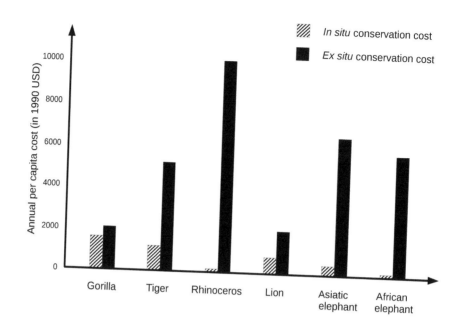

(b) Cost of conserving different priority species.

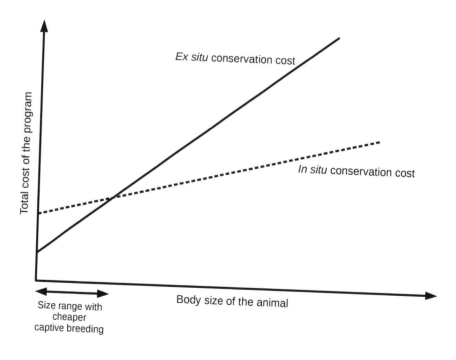

(c) Cost of conservation with respect to the body size of the animal. *Ex-situ* conservation may be more cost-effective for small body-sized animals.

Figure 10.2: Comparative data between *in-situ* and *ex-situ* conservation options. Data source: [Balmford et al., 1995].

10.2.1 *Ex-situ* conservation

Ex-situ conservation normally proceeds in three stages. First, areas with suitable conditions (such as access, availability of water, gentle slope, etc.) are selected and appropriate facilities (such as enclosures, paths, animal clinics, etc.) are created. In the second step, species are moved into these designated areas — either from the wild, or from other *ex-situ* conservation facilities — for their survival and breeding. In this step, the organisms are fed and taken care of. The third, optional step, is to release the species back into their natural habitats once their numbers have gone up. Of course, there are details and procedures involved in each of these steps.

In the case of animal facilities such as zoos and rescue centres, the process begins with a master plan [Fig. 10.3a]. This plan, which may need to be approved by national nodal agencies, provides details about various aspects of the creation and management of the zoo. It is often accompanied by the stud-books for various animals [Fig. 10.3b] — needed to ascertain the provenance and lineage of individuals so as to prevent inbreeding and genetic diseases.

The animals may come from the wild (generally rescued animals or those caught specifically for captive breeding programs), from other zoos (through transfer, purchase, or loan), or may be born in the facility [Fig. 10.3c,d]. The animals are kept in enclosures that satisfy their requirements [Fig. 10.3e,f] of space, shade, cover, and food [Fig. 10.3g,h]. The enclosures often also have means of providing behavioural enrichment to animals [Fig. 10.3i] so that they are occupied and do not show stereotypic behaviours. Animals with specific requirements such as basking are provided appropriate facilities in their enclosures [Fig. 10.3j].

Zoos are required to maintain extensive records [Fig. 10.3k] regarding health of the animals, their feeding schedule, training activities, visitor data, etc. They often have good veterinary facilities on site [Fig. 10.3l], or are associated with nearby veterinary hospitals or colleges. Zoo infrastructure also includes various visitor amenities [Fig. 10.3m] to cater to the needs of the visitors and researchers.

In the case of plants, the creation of *ex-situ* conservation stands involves site selection and making decisions about the size and handling of plantation, sampling of source population to select the individuals to be brought in, collection of seeds of good quality [Fig. 10.3n], raising of plants, establishment of plantation, and various management operations such as weeding, irrigation, and fertilisation.

Botanical gardens are often created in areas that can provide a variety of habitats [Fig. 10.3o], including specialised habitats for priority species [Fig. 10.3p]. They often support a large biodiversity [Fig. 10.3q], and thus act as great educational centres [Fig. 10.3r] for students, researchers, and general public [Fig. 10.3s] — providing not just scientific information, but also helping to create a strong public opinion towards conservation.

Cryo preservation facilities also act as important *ex-situ* conservation areas [Fig. 10.3t] where a variety of biological materials — from seeds of plants to sperms, eggs, and zygotes of animals — can be kept and preserved. The samples are often kept in deep freezers [Fig. 10.3u] and are used for artificial insemination, gamete transfer, embryo transfer, and various studies [Fig. 10.3v] — on aspects like the earliest occurrence of diseases and changes in the genetic makeup of animals over time. Cryo-preservation of sperms and ova also provide hope to use future technologies for wildlife conservation.

Ex-situ conservation offers several advantages. Keeping animals in a small, artificial location permits excellent control of variables such as climate, diseases, diet, etc. The facility furnishes the opportunity to observe the animals closely — permitting a better understanding about the species characteristics, behaviours, and requirements. This often leads to insights into the causes that are pushing the species towards extinction. At the same time, *ex-situ* facilities allow intensive interventions such as *in-vitro* fertilisation, embryo transfer, etc., which may be urgently needed to save the species from extinction.

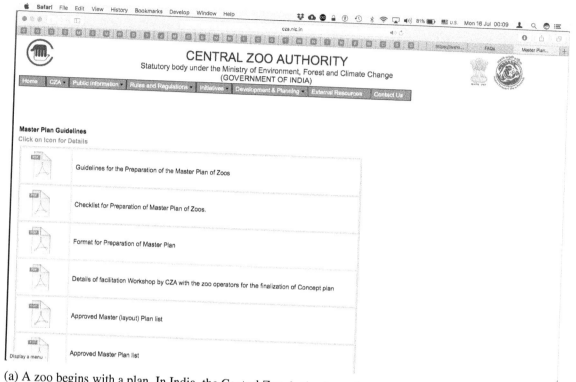

(a) A zoo begins with a plan. In India, the Central Zoo Authority makes regulations, approves master plans, coordinates and aids implementation of conservation breeding in zoos.

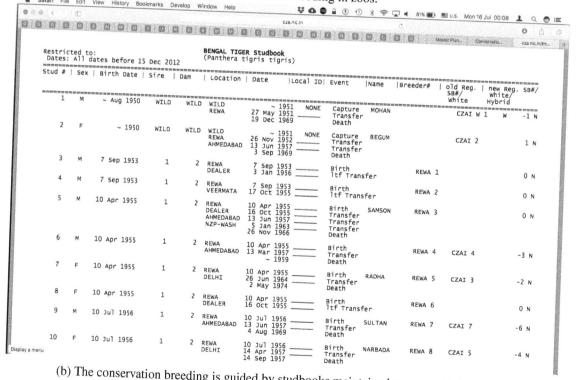

(b) The conservation breeding is guided by studbooks maintained to prevent inbreeding.

(c) The image shows turtle hatching site at the Turtle rescue facility, Dwarka.

(d) After they hatch, the animals are kept in controlled conditions till they are mature enough to be released.

(e) Zoos provide cosy environments for their residents.

(f) A view of sloth bear enclosure in the Agra Bear Rescue Facility.

(g) Provisioning of adequate, safe, nutritious food is a big task in zoos.

(h) At times, the zoos may have on-site animal rearing facilities to provide food for the inmates. When the food is brought from outside, quality checks are especially important.

(i) The animals are provided with behaviour enrichment to keep them occupied and to prevent stereotypical behaviours. These enrichments can be in the form of toys, puzzles, jumping bars, etc.

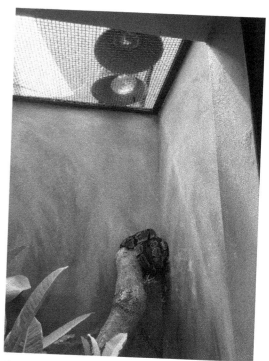

(j) Knowledge of animal behaviours can aid management. Here we observe a heating lamp in the snake enclosure at Johannesburg Zoo, provided to aid the natural basking behaviour of snakes.

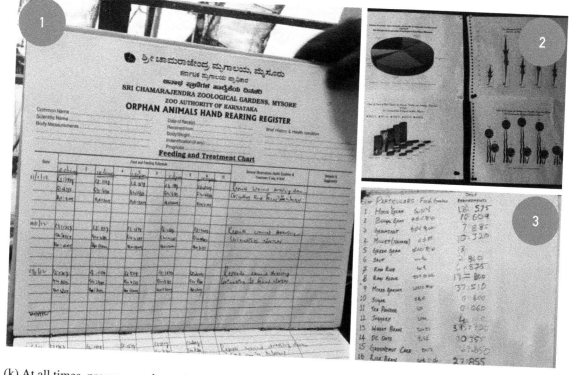

(k) At all times, proper records need to be maintained regarding health of inmates, feeding schedule, training activities, visitor data, etc.

(l) Zoos also have good veterinary facilities to cater to the health of the inmates.

(m) Since zoos act as means of outreach and information, adequate provisioning of visitor facilities is essential.

(n) Seed banks and botanical gardens must ensure collection and usage of fully ripened, healthy seeds.

(o) A large botanical garden, such as the Kirstenbosch botanical garden in Cape Town, is ideally sited such that a variety of habitats — from plains to mountainous — are available to raise a variety of plants.

(p) Rearing of plants such as orchids and cacti requires specific environments.

(q) Botanical gardens not only support plants, but also a variety of animals and birds.

(r) Botanical gardens provide outreach and educational opportunities, aided by the use of captivating information boards.

(s) *Ex-situ* conservation sites provide close access to nature for families.

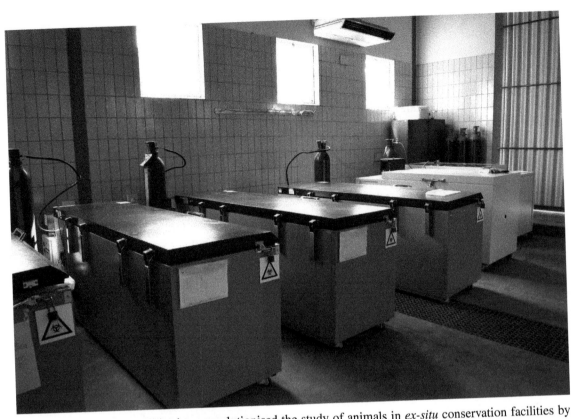

(t) Cryo preservation facilities have revolutionised the study of animals in *ex-situ* conservation facilities by providing a means to keep large number of samples for very long periods.

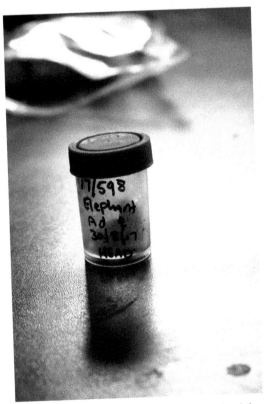

(u) The samples are kept in deep freezers, often at −18 °C or lower temperatures.

(v) A cryo-preserved sample in a sample vial.

Figure 10.3: Examples of *ex-situ* conservation areas and activities.

Ex-situ conservation also has several limitations on its own. It does not do anything to prevent the loss of habitat. If *ex-situ* conservation is not accompanied by protection and conservation of habitats, it may lead to a situation where the captive animals do not have a home to go back to — a place where they can be released once their populations have stabilised — thus converting them into mere artefacts that do not get to play any ecological role. It is costly, and being expensive and more resource-intensive, can be planned for only a few species at a time, typically those species with small body sizes. Wild behaviours frequently get lost in the artificial environments of the facility, with the result that captive-bred and raised individuals find it difficult to thrive when they are reintroduced back into their natural habitat. If not planned properly, *ex-situ* conservation also increases the chances of inbreeding once the few animals that are kept together in the facility become relatives of each other.

Genetically, *ex-situ* conservation may have several unintended implications [Hamilton, 1994], such as

1. Stochastic sampling of alleles: When individuals (or seeds) are collected from the natural environment to be kept in the *ex-situ* facility, the sampling randomly selects some alleles while discarding other alleles in a stochastic (random) manner. Thus, some amount of natural variation always gets lost in the sampling process, similar to the effect of a population bottleneck. This needs to be compensated — by extensive sampling from different geographical locations and meticulous collection of naturally occurring variations. If not done rigorously, there is a good chance that the individuals collected do not represent the variations naturally occurring in the species.

2. Erosion of genetic variation: The natural habitat often provides a dynamic environment — one that is constantly changing. Sometimes there is an ample supply of food; at other times there is a famine. Sometimes it is too hot, at other times it is freezing cold. Spells of rains may be followed by months of drought. Thus, the only constant in a natural habitat is change. To adjust to the changing conditions, the organisms develop variations. Thus, while some organisms are better adapted to cold, others are better tolerant of heat. Some are drought resistant, while others are resistant to water logging. This variation ensures that the population survives even when the conditions become extreme. And regular changes in the environment act as selection pressures to maintain these variations.

   The conditions in the *ex-situ* conservation facility, on the other hand, are often much more consistent and regulated. The individuals are kept in climate-controlled, clean conditions with abundance of food and care. In consistent conditions, there is no selection pressure to keep the genetic variations in the population — especially those that play a role in adapting the population to changes. This leads to an erosion of genetic variations — the variations get reduced with each subsequent generation, and the individuals become more and more like each other.

3. Pleiotropic effects/Genetic correlations: Some genes depict the phenomenon of pleiotropy (Greek *pleion* = more, *tropos* = way) — they influence multiple, at times unrelated, phenotypic traits. The same gene may increase stability during storage but decrease number of seeds produced. In such cases, the selection of plants producing seeds with better storage stability — that can safely be stored for long durations — will also result in selection of plants that produce less number of seeds. This would be antagonistic to the objectives of reintroduction — for when the seeds are later planted in the natural habitat, we would ideally prefer the plants to produce a large number of seeds themselves.

4. Genotype-environment interactions: Phenotypes result from an interaction of genotype with the environment. Since the environment in an *ex-situ* conservation facility is often very different from the environment in the natural habitat, the phenotypes presented by the same

genotype in the two environments may be very different. Thus, plants selected for producing large number of seeds in a seed bank may produce very few seeds in the natural habitat. This may lead to selection of non-optimal genotypes — those are not the best-suited ones for later re-introduction.

10.2.2 In-situ conservation

In-situ conservation is done by designating areas in the natural habitat as reserves, sanctuaries, national parks, or protected areas, followed by ecological monitoring and interventions (active management) as and when required. Often legislations are required to maintain these areas as protected areas. The operations required for the management of the reserves have already been detailed in chapter 9. In this chapter, we discuss the area selection and design of the reserves.

Traditional ways of creating reserves selected areas based on the whims and fancies of the reserve creator. Often these were beautiful areas — with lush green mountains, lakes, or beaches. Dachigam National Park in Srinagar is one such beautiful spot. In other times, areas with high species diversity, such as the Silent Valley in Kerala were selected to be designated as National Parks. Often these were the areas that had served as hunting grounds or *shikargahs* for the local rulers. Areas harbouring unique animals were also often made into reserves. For example, the Gir National Park in Sasan, Gujarat has the only remaining population of Asiatic lions.

The rapid loss of biodiversity has necessitated creation of protected areas that are best suited to the cause of conservation — often with minimum resources. We cannot just depend on whims and fancies — the process needs to be rigorously scientific and economically defensible when we make use of public money.

Thus the modern processes aim to create reserves in ways such that maximum conservation can be achieved with minimum resources. To this end, we try to select areas with

1. high species richness, and/or
2. high species endemism, and/or
3. high number of species under threat,

with the best suited areas being those that have all the three characteristics together.

Selection of areas with high species richness allows for conservation of a large number of species with the same amount of resources and effort. To this end, we make use of maps of global species richness [Fig. 10.4a,b]. These maps may represent all the biodiversity, or biodiversity in specific clades — such as mammalian or amphibian richness — which may be used as a proxy for the richness of all biodiversity.

Endemism is the characteristic of a species being extremely localised to a single geographical location. Such species are only found in these restricted areas, and nowhere else. Typically these areas have specialised conditions that are not found in any other area. In other cases, species that have suffered severe habitat loss and poaching — such as rhinoceros or asiatic lions — are now found in only certain patches. These are thus the only areas left in the former extended range of the species. If these areas get lost, damaged, or destroyed, the species will get exterminated — for ever. Thus, protection of the contemporary range of these species is crucial for their survival and existence. Areas with high species endemism support many such species that are not found anywhere else. Thus, their protection is pivotal for the continued existence and sustenance of several species, making them priority areas to be declared as reserves.

The number of species under threat can similarly be used as a characteristic to select priority areas. If a large number of threatened species live in a particular place, protection of this place becomes important if we wish to protect these species from extinction. Such areas have been mapped for aiding decision-making [Fig. 10.4c]. On the basis of the level of threat, we can divide areas into three categories — called the threat triage:

1. Category 1: Areas facing very high levels of threat (say, due to imminent extensive deforestation to make way for agriculture or mining) — these areas may already be a lost cause, for there will be a huge public resistance (since people have already occupied these areas), and these areas will (probably) be completely destroyed in the time it takes to create the reserve! The work to create reserves should ideally begin much before the threat reaches very high levels. If these areas are crucial for conservation — such as tropical rainforests and coral reefs — a more prudent approach is to create reserves with a 'buffer' zone (areas that will not be included in the reserve — typically that portion of habitat that is already lost and occupied by humans, and whose incorporation will create huge resentment, delaying the whole process) so that maximum areas can be conserved as quickly as possible.

2. Category 2: Areas facing a medium level of threat — this is where most of the focus should be, for conversion of these areas into reserves will probably have a large impact on conservation! Also, if these areas are not declared reserves now, in future — as the threat levels go up — they will move to category 1.

3. Category 3: Areas facing very low levels of threat (say, because they are so far off and uninhabited by humans (think uninhabited islands) that there is virtually no threat of habitat loss) — these areas may be left for now, for there is little impact of conversion of these areas into reserves; the effort and resources may be better utilised elsewhere, to protect category 2 areas with medium level of threat.

These three criteria — richness, endemism and threat — can be combined to form a composite picture. We define biodiversity hotspots are areas with

1. high species richness,

2. high degree of endemism, and

3. high degrees of threat.

A map of the biodiversity hotspots of the world is depicted in figure 2.5. We may also make use of composite indices such as the Biodiversity Intactness Index (BII) [Fig. 10.4d] to help locate areas that still have a considerable biodiversity left intact which can be saved through the creation of reserves. BII is defined as "the average abundance of originally present species across a broad range of species, relative to abundance in undisturbed habitat [Newbold et al., 2016]." Areas with low BII are often already a lost cause for conservation.

Historically, wildlife reserves were often created in areas that did not have any other utility for humans — areas that were considered 'wastelands.' Often this occurred because the areas were too hot, too cold, too infested with mosquitoes, too far off or too arid. This becomes clear when we analyse the current reserves. The height map of the Khangchendzonga National Park in Sikkim [Fig. 10.4e], for instance, shows that many locations are so high that they are not of much use to humans. Similarly, the Desert National Park in Rajasthan [Fig. 10.6t] is so dry that there is hardly any utility for humans. This is the reason why these areas could be 'spared' for wildlife conservation.

While this approach — of creating wildlife reserves in areas selected on the basis of their *non-utility* for humans — does protect the small number of species that live in harsh habitats, we need to appreciate that these are not the optimum areas for the sustenance of wildlife. Wildlife had also existed in areas having amiable climate and access to food and water — before these areas were diverted for exclusive use by humans. As a result, the condition of species that lived in

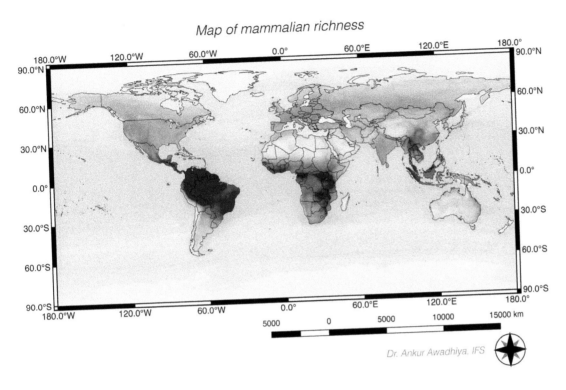

(a) The richness of several species is now well documented. The map shows global mammalian richness.

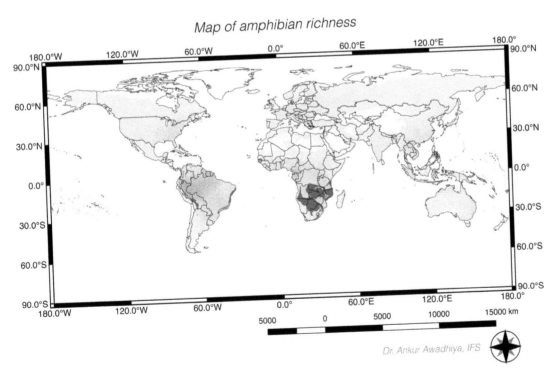

(b) During area selection, adequate attention must also be given to non-mammalian forms of biodiversity, especially amphibians and other herpetofauna. The map shows global amphibian richness.

Map of threatened mammalian richness

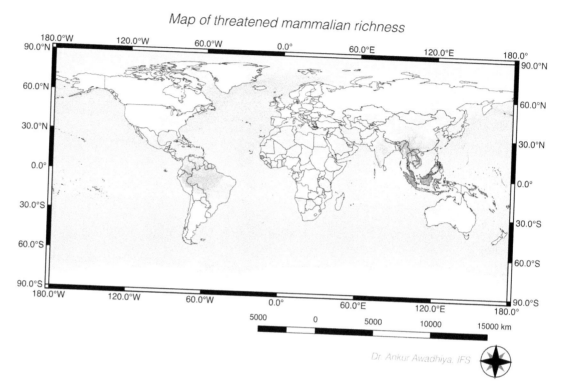

(c) If an area has more number of species under threat, and something can *still* be done about that, it should be done on a priority basis. The map shows the global *threatened* mammalian richness.

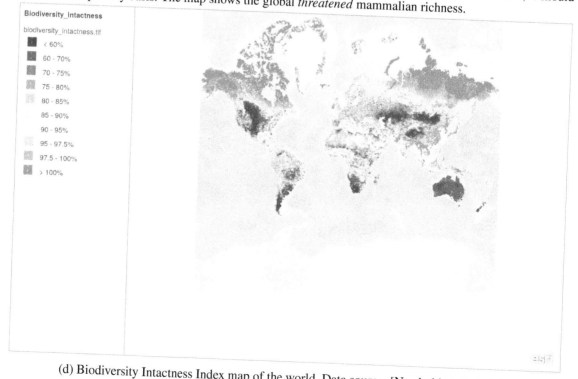

(d) Biodiversity Intactness Index map of the world. Data source: [Newbold et al., 2016].

Height map of Khangchendzonga National Park

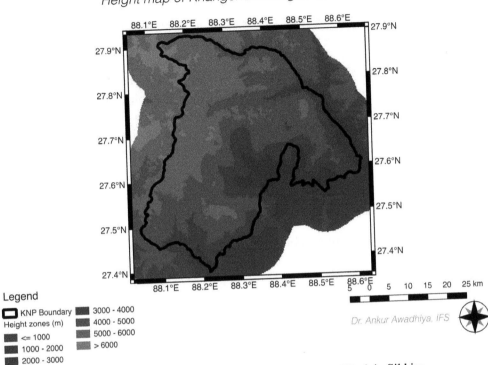

Legend

KNP Boundary
Height zones (m)

<= 1000
1000 - 2000
2000 - 3000
3000 - 4000
4000 - 5000
5000 - 6000
> 6000

Dr. Ankur Awadhiya, IFS

(e) The height map of Khangchendzonga National Park in Sikkim.

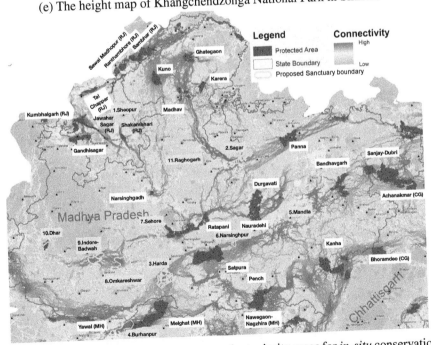

(f) Through an application of these principles, we may select priority areas for *in-situ* conservation. The figure represents suitable locations for new sanctuaries in Madhya Pradesh — those that have high biodiversity, moderate levels of threat, can fill in the gaps, and act as corridors for supporting animal connectivity.

Figure 10.4: Selection of area for *in-situ* conservation.

the now-diverted areas is especially precarious — and we need to do more for them. Such areas can be selected through *gap analysis* — an approach that tries to identify holes in the existing network of protected areas. Creating some protected areas in these human-dominated areas — say by purchasing agricultural land for conversion into wildlife reserves — may fill the gap, allowing a different set of species to thrive.

Considering all these criteria together can help us to scientifically identify those locations where creation of wildlife reserves will bring the maximum benefit with the minimal resources [Fig. 10.4f].

Once the suitable areas have been identified, the next step is to create the reserves, while being mindful of the *principles of reserve design* [Fig. 10.5]. These principles aim at enhancing connectivity within and between reserves, to convert small populations into a large population and reduce the impacts of the small population paradigm. They also try to reduce the biotic pressure from human activities by minimising the perimeter of the reserve, and thus, the zone of influence of human activities. The six principles of reserve design are:

1. Big is better than small [Fig. 10.5a]: Given a choice, we should try to create as large a reserve as possible. This is because of several reasons:

   (a) Bigger areas have more number and variety of habitats, resulting in higher species diversity.

   (b) Bigger areas are more secure and easier to manage (per unit area). This is because

      i. Larger populations can be sustained in larger areas, making them less susceptible to extinction, since small population dynamics do not play a role.

      ii. Larger areas result in smaller perimeter: area ratios. The greatest cost in managing a reserve is in protecting the reserve — keeping intruders such as poachers and livestock out. This protection has to be done at the boundary of the reserve. On the other hand, the benefit comes from the species that reside in the 'area' of the reserve — greater the area, greater is the benefit. Thus, a smaller perimeter: area ratio means getting larger benefits with lesser costs.

      iii. Larger areas are less vulnerable to catastrophes since most catastrophes will not impact the whole area. If a small habitat is completely engulfed by fire, there will be very less, if any, individuals left to re-populate the area. On the other hand, a larger area provides insurance through its sheer size. There is minimal chance of any catastrophe impacting the large area completely, and the individuals that survive will re-populate the reserve themselves — without the need to translocate individuals artificially. In this way, larger reserves have larger resilience to many catastrophes.

2. One big is better than several small of same total area [Fig. 10.5b] — because the animals can freely move from one place to another, and there is less pressure from the surroundings through reduction of perimeter.

3. Prefer closer reserves, to minimise isolation [Fig. 10.5c] — closer reserves permit greater and easier movement of animals from one reserve to another, without having to pass through large human-dominated areas. This converts several small populations into a large population, reducing the factors of the small population paradigm.

4. Prefer clusters, since a cluster arrangement permits greater movement than a linear arrangement [Fig. 10.5d].

5. Prefer making circular reserves — they suffer less biotic pressure due to their small perimeter: area ratio [Fig. 10.5e,g].

(a) Big is better than small.

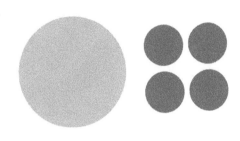

(b) One big is better than several small of same total area.

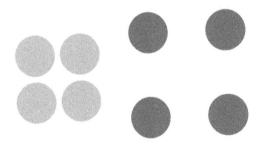

(c) Closer reserves minimise isolation.

(d) Cluster permits more movement than linear.

(e) Circular reserves have less biotic pressure.

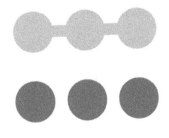

(f) Connection is important.

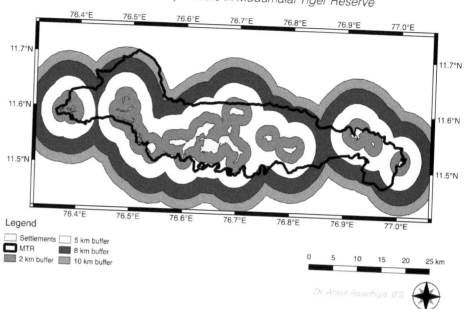

(g) Map of biotic pressure in Mudumalai Tiger Reserve, represented by buffers around settlements. A 5 km buffer covers *most* of the reserve, while a 10 km buffer covers *all* of the reserve.

Figure 10.5: Principles of reserve design.

6. Connection is important: Always strive to maintain/enhance connectivity [Fig. 10.5f] within and between reserves.

Some examples of *in-situ* conservation areas are depicted in figure 10.6. *In-situ* conservation offers several advantages:

1. Species continue to live in their natural environment — thus their natural behaviours are maintained and conserved.

2. It is less disruptive and more inexpensive than *ex-situ* conservation.

3. Protection of the natural habitat of target wildlife provides collateral protection to several species of organisms, without any extra effort.

4. Most *ex-situ* conservation is done with the target of releasing the animals back to their natural habitat, someday. *In-situ* conservation sites provide suitable areas for such releases.

5. Reserves double as places for scientific studies, tourism, and public awareness, besides providing employment to local communities.

6. Reserves also act as places providing several economic values [Section 1.3] and ecosystem services — such as water purification, carbon sequestration and storage, and groundwater recharge — to the community and the nation.

However, *in-situ* conservation also has certain limitations:

1. It requires large areas that have to be earmarked for conservation — and land often has a large opportunity cost since it can also be put to multiple other uses.

(a) The landscape of Dachigam National Park, Jammu and Kashmir.

(b) A Himalayan tahr in the Kedarnath Wildlife Sanctuary, Uttarakhand.

(c) Elephant with calf in Mudumalai Tiger Reserve, Tamil Nadu.

(d) Nilgiri tahr in Mukurthi Wildlife Sanctuary, Tamil Nadu.

(e) The landscape of Nalsarovar Bird Sanctuary in Gujarat.

(f) The river Chambal in National Chambal Sanctuary in Madhya Pradesh supports a considerable population of gharial and Ganges river dolphin.

(g) The landscape of Bharatpur Bird Sanctuary in Rajasthan provides a wetland in the desert state and supports a large bird diversity.

(h) The landscape of Khijadiya Bird Sanctuary in Gujarat provides a wetland in an otherwise parched area.

(i) The Fudam Bird Sanctuary is in the Union Territory of Diu. Note the large wetlands.

(j) Wetlands also dominate the landscape of Nalbana Bird Sanctuary in the state of Odisha.

(k) Marine species are provided protection in the Gulf of Mannar Marine National Park in Tamil Nadu.

(l) The Mahatma Gandhi Marine National Park in Andaman Islands supports many endemic species.

(m) The Marine National Park in Gujarat permits observation of several sea organisms during low tide.

(n) The Buxa Tiger Reserve in West Bengal provides an altitudinal variation from 60 m above sea level to 1,750 m above sea level, thus supporting a large biodiversity.

(o) The Ranthambhore Tiger Reserve in Rajasthan supports a large tiger population in an arid landscape.

(p) The Kanha Tiger Reserve in Madhya Pradesh has a large population of barasingha (*Cervus duavcelli branderi*) adapted to living on hard ground.

(q) The Gir National Park in Gujarat has the last remaining population of the Asiatic lion (*Panthera leo*).

(r) The Blackbuck National Park in Gujarat has vast grasslands supporting a large population of blackbuck (*Antilope cervicapra*).

(s) The Indian Wild Ass Sanctuary in Gujarat protects the endemic Indian Wild Ass (*Equus hemionus khur*).

(t) The Desert National Park in Rajasthan protects a variety of desert organisms.

Figure 10.6: Examples of *in-situ* conservation areas.

2. It allows only less intensive protection and management — thus the areas may be encroached upon, and animals may get poached. This is especially because it is very difficult to completely protect and manage large areas at all times.

3. There is a constant threat of diseases and disasters. Being large areas, they do not permit extensive mitigation of threats, as is possible in *ex-situ* conservation — such as through provisioning of intensive veterinary care.

4. Often, *in-situ* conservation reserves have large establishment costs, since many people have to be employed for protection and management of large areas. While the cost per unit area is less, the overall costs may nonetheless be substantially high.

The difficult choice, in most cases, is not choosing between *ex-situ* conservation and *in-situ* conservation — for we need *both of them* for effective conservation. The difficulty often lies in choosing between conservation (which gives resources and benefits slowly, but perpetually) and unsustainable development (which provides large short-term gains at the cost of large long-term harm). Once the intention is clear, both these routes — *ex-situ* conservation and *in-situ* conservation — converge on the same objective — *conservation*. They complement each other, and are never competitors!

CHAPTER 11

Emerging aspects of wildlife management

Wildlife management is quickly evolving with time. Today's wildlife management is data-hungry and data-centric — requiring *lots* of data. A major chunk of this data is now being collected using remote sensing — especially to identify locations requiring management interventions such as removal of weeds, fire preparations, or provisioning of drinking water. We also require data to justify the creation and/or continuation of wildlife conservation facilities. This is especially so because increase in human population and affluence has skyrocketed the cost of land. Land can be put to multiple uses — agriculture, setting up of industries, constructing houses and schools, mining, etc. Such uses of land increase prosperity and affluence of people, and so in democratic countries it is becoming more and more difficult to convince policy makers to spare land for conservation causes such as wildlife reserves. Mere appealing to their hearts will not work — we need to demonstrate that setting up of a wildlife reserve also has numerous monetary advantages — such as through ecosystem benefits. But to convince the policymakers, these ecosystem benefits will have to be calculated in terms of their dollar (or currency) value — to permit cost-benefit analyses. Only when the benefits — in monetary terms — exceed the costs — including the opportunity costs in terms of benefits forgone from other *developmental* activities — will the policy makers agree to create reserves.

In this chapter, we discuss these emerging aspects.

11.1 PHOTOGRAMMETRY, REMOTE SENSING AND GIS

Photogrammetry is the "science and technology of obtaining spatial measurements and other geometrically reliable derived products from photographs." In other words, it is the technique of using photographs to make computations and measurements. Often this involves converting 2-d images into 3-d models so that measurements in any direction can be made reliably [Fig. 11.1]. It is based on the principle: "Triangulation permits depth perception," meaning that we can use principles of triangulation to derive 3-d information (about depths) from 2-d photographs.

Taking photographs is a form of remote sensing — sensing information about an object in a remote manner, without making a physical contact. There are three components to remote sensing:

1. an object about which some information is needed,

2. a recording device — such as a camera, and

3. information carrying energy waves — light, radio waves, etc.

In photography, the light rays bring information from the object to the camera, which records this information. The complete process can be written in eight steps:

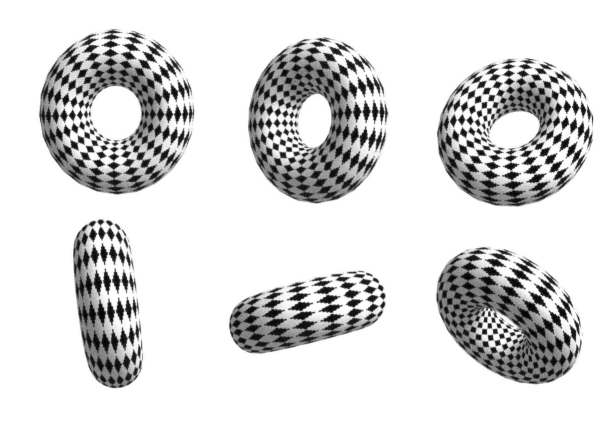

(a) Photogrammetry uses triangulation for computation of depths.

(b) A portion of the meadows of Kanha Tiger Reserve constructed in 3-d using photogrammetry. Z exaggeration = 20×.

Figure 11.1: Principle and practice of photogrammetry.

Energy generation (own energy of a luminous object/energy from the Sun) → Transmission of energy to the object of interest → Interaction of energy with the object of interest → Propagation of reflected/emitted energy from the object to the sensor through the atmosphere → Detection of reflected/emitted energy by the sensor → Conversion of detected energy into data → Extraction of information from data → Conversion of information into a presentable form (such as a printed photograph or a JPEG image).

In photogrammetry, once the photographs have been taken from multiple locations (at least two locations), 'lines of sight' are drawn from each camera location to points on the object. These 'lines of sight' are then mathematically intersected to get 3-d coordinates of the points of interest. Since a lot of data gets generated, we often make use of Geographic Information System (GIS). GIS is a system consisting of hardware (equipment such as computers), software (code) and humanware (people who work with the data) that is designed to capture, store, manipulate, analyse, manage and present different kinds of geographical data. It is an integrated system to process and analyse geographical data, including those received through remote sensing.

Modern GIS technologies handle data in a digital format. The data is represented in the form of layered stacks that are interconnected and can be queried.

To facilitate decision-making, data from a plethora of sources can be integrated together in a GIS. These data can be satellite data, photographs, scanned copies of maps, data from old documents (earlier reports), toposheets, GPS coordinates from field visits, field surveys, transacts and their description forms, cruises, etc. These data are georeferenced to permit seamless integration with other geographical data [Fig. 11.2].

The information generated through GIS can be metric information (that can be measured — such as number of trees per unit area) [Fig. 11.2a] or interpretive information (used to interpret the actual situation — such as discerning the areas facing drought or the extent of water available in a water body) [Fig. 11.2b].

Common applications of GIS are planning, monitoring land use changes, understanding topography and hydrography, exploration, and reconnaissance. These are aided through the generation of thematic maps, orthophotos, and digital elevation models, such as shown in figure 11.1b.

Together with photography using sunlight (a form of passive remote sensing — not requiring energy for the generation of probing radiation), these days we also make use of active remote sensing (using energy for the generation of probing radiation) methods such as RADAR (radio detection and ranging) and LiDAR (light detection and ranging). RADAR and LiDAR are similar techniques — the main difference being the radiation used in the process — radio waves versus light waves. The wide and easy availability of light weight LiDAR equipment has stimulated its extensive usage for forestry applications — so let us discuss it in more detail.

LiDAR is a method of Air-borne Laser Scanning (ALS) developed in 1960 by Hughes Aircraft, Inc. It shoots LASER beams — which are monochromatic (of a single wavelength) and extremely directional (meaning that they exhibit very little divergence over long distances) beams — to the targets (here, land, water and vegetation). These LASER beams interact with the target, and then get reflected back. The reflected beams are detected and measured.

The exact position and direction of the LiDAR equipment is measured using precise equipment such as differential GPS and inertial measurement units. The distance to the target is calculated by the time taken by the LASER beam to come back:

$$d = c \times \frac{t}{2}$$

where d = distance of the target from the LASER source,

c = speed of light (299,792,458 m/s), and

t = time taken by the LASER beam to come back to the LiDAR equipment.

By keeping track of angles, we can make a 3-d scan of the target. In this way, for LiDAR to function, we require four components:

(a) Georeferencing permits overlaying of drone imagery over satellite imagery. Drones — since they fly at lower altitudes — can provide much higher resolution imageries than satellites. The use of georeferencing permits precise overlaying. Notice how the roads are exactly overlaid in the two imageries. The picture can be used to count the number of trees planted on the ground — an example of metric use of photogrammetry.

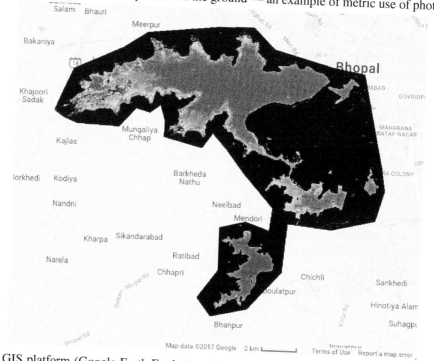

(b) Use of a GIS platform (Google Earth Engine) to discern the maximum availability of water in a water body is an example of an interpretive use of photogrammetry.

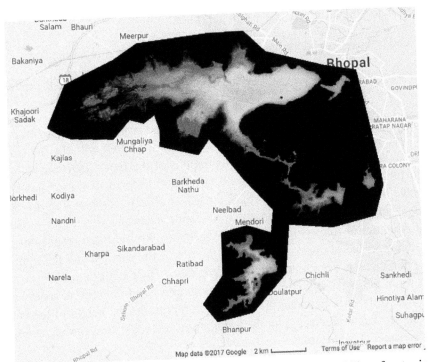

(c) Use of a GIS platform (Google Earth Engine) to discern the average availability of water in a water body is an example of an interpretive use of photogrammetry.

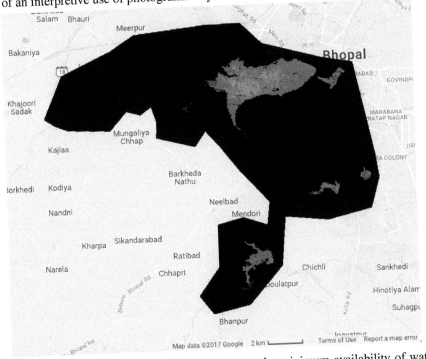

(d) Use of a GIS platform (Google Earth Engine) to discern the minimum availability of water in a water body is an example of an interpretive use of photogrammetry.

Figure 11.2: Metric and interpretive use of GIS information.

1. a LASER source — to generate the beam,

2. positioning and navigation systems — to record the exact location of the LASER source and the direction to which the LASER beam gets shot,

3. scanner and optics — to shoot the LASER beam to different directions, and

4. photodetector and receiver electronics — to receive, measure and record the reflected LASER beams.

LiDAR can be deployed from various heights — from the ground level to high up in the space. The deployment at great heights — such as the integration of LiDAR with the International Space Station — permits a synoptic viewing of various regions of the Earth, together with capturing of redundant data for improved accuracy [Fig. 11.3a]. When used to monitor forests and wildlife habitats, the waveform of the returned LASER pulses — after interaction with trees, ground and water surface — can provide information about the land, water and vegetation — even regarding the height and structure of trees [Fig. 11.3b]. For instance, the light that reaches back first is the one that has been reflected at upper levels — by the top of the canopy, and the light that reaches back last is the one that has been reflected at lower levels — by the ground. Thus, we can plot two curves — one representing the ground surface (topography) and the second representing the top of the canopy. The difference between the two curves represents the height of the canopy. In this way, LiDAR can be used to precisely plot the topography and vegetation canopy height at different locations on the planet [Fig. 11.3c].

A more detailed analysis of the returned waveform can also provide information about different sections of the tree, from the ground level to the top [Fig. 11.3d]. In this way, LiDAR can be used to represent 3-d sections of the forest, with information about the topography and the biomass at different heights [Fig. 11.3e]. This can also be substantiated by LiDAR data from ground level [Fig. 11.3f], to arrive at high resolution vegetation and carbon storage maps [Fig. 11.3g] for various locations on the planet. These cannot only be used for assessing the loss of habitats and for planning various habitat management operations, but also to arrive at economic valuation of different forests and their ecosystem services.

11.2 ECONOMIC VALUATION OF PROTECTED AREAS

We had noted in chapter 1 that wildlife is a biotic, renewable, actual (sometimes potential) resource providing several economic values [Fig 1.3] — including direct consumptive values such as timber and honey, direct non-consumptive values such as education and tourism, indirect values and several non-use values such as existence, altruistic and bequest values. If we can compute the dollar (or monetary) values of all the benefits that are being provided by wildlife, we can use that value to make decisions. These decisions include setting up of new wildlife reserves for the benefit of the local community, setting up of zoos and aquaria for education and research, and diversion of existing wildlife habitats for activities such as construction and mining. In this section, we shall examine how best we can compute the value of wildlife resources. The process may similarly be used for any other resource too.

We begin by making a list of the services that are being provided by wildlife in their habitat. These services include

1. provisioning services such as food, medicines, etc.,

2. regulating services including climate regulation, biological control of pest population, etc.,

3. supporting services namely soil formation, nutrient cycling, etc., and

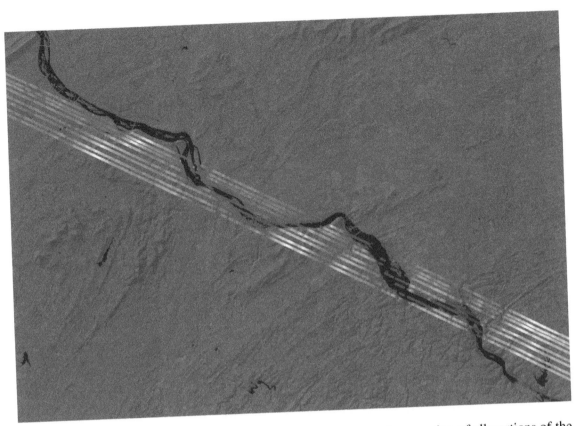

(a) Installation of LiDAR at the International Space Station permits scanning of all portions of the Earth to arrive at a synoptic understanding of various forests. Image source: [NASA, 2019b].

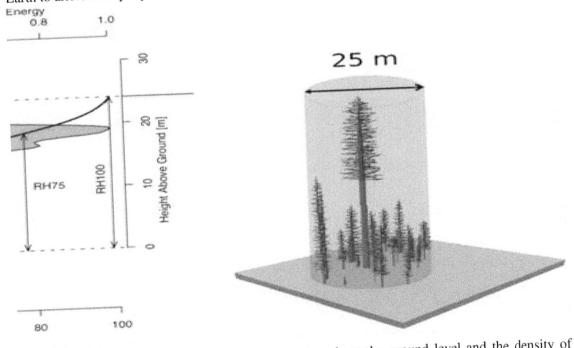

(b) The returning wave profile provides information about the ground level and the density of biomass at different heights above the ground. This can be used to reconstruct the trees in various forests. Image source: [NASA, 2020a].

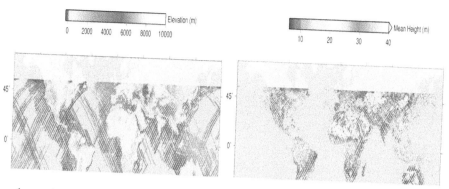

(c) Topography and vegetation height maps can be generated through LiDAR data. Image source: [NASA, 2019a].

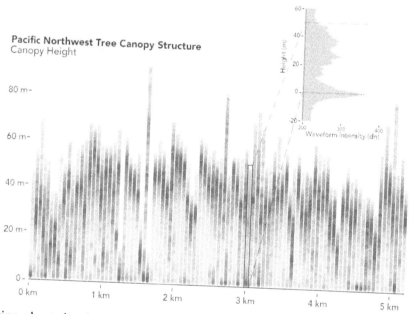

(d) Information about the density of biomass at different heights of tree can be discerned from LiDAR data. Image source: [NASA, 2020b].

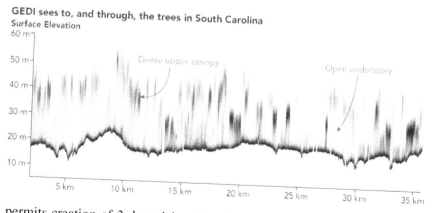

(e) LiDAR permits creation of 3-d models of the forests, representing the amount of biomass at each point in the 3-d space. Image source: [NASA, 2020a].

(f) Application of LiDAR from ground level in the form of Terrestrial Laser Scanner can provide high-resolution information about forest structure. Image source: [NASA, 2017].

(g) High-resolution carbon storage maps generated using LiDAR can be used for planning of habitat management operations and for economic valuation of forests and ecosystem services provided by them. Image source: [NASA, 2012].

Figure 11.3: Use of LiDAR for gaining information about forests.

4. cultural services — e.g. recreational, educational, religious services, etc.

Some of these services can be monetarily valued by looking at their market prices. For instance, we can compute the amount of timber in a forest and multiply that value with the current market price of timber — to arrive at the monetary value of timber in the wildlife habitat. But such a method will only value a minuscule portion of the services being provided by wildlife — since not all services are sold in the market. After all, we cannot compute the value of an elephant by calculating the market value of its tusks, skin, flesh, and bones — a living elephant has several cultural, religious and recreational values that a dead elephant doesn't have. After all, so many people visit elephant reserves to watch elephants roaming, bathing, and feeding, but hardly anybody would visit an elephant reserve sporting only dead elephants! In such cases, how can we put a monetary value on the life of an elephant? Similarly, how do we calculate the monetary value of the cultural significance of a sacred grove? Or of the joy obtained on sighting a tiger?

One easy way out is to say that we will only compute the values of those things that have a market price. And this is what has been done in the field for quite a while, resulting in grossly minimised values of wildlife and their habitats. Often this was done not just because we did not know how to compute, but also to favour the industries that aimed at diverting wildlife and their habitats — a *mala fide* intentional disvaluation of wildlife and their habitats for short-term profits. Surely we need to do better. So what options do we have?

11.2.1 Methods of valuation

Currently, there are three accepted approaches for valuation:

1. market prices/revealed willingness to pay, including

   (a) market price method,
   (b) hedonic pricing method, and
   (c) travel cost method,

2. circumstantial evidence/imputed willingness to pay, such as

   (a) replacement/substitute cost method and
   (b) damage cost avoided method, and

3. surveys/expressed willingness to pay, such as

   (a) contingent valuation method.

The "revealed willingness to pay" methods opt for cases where people have disclosed their willingness to pay for a service by actually paying for the service. Thus, if the market price for timber is X dollars per cubic feet, it means that the *value* of 1 cubic feet of timber, as *revealed* by the buyers and sellers in the market, is X dollars. If a forest has Y million cubic feet of timber, then the value of timber is the forest is $X \times Y$ million dollars.

For all the *goods that are traded in the market* — timber, honey, medicinal herbs, fodder, hides, bushmeat, etc. — we compute the amount of each product and multiply it with the prevailing market price of that product, giving the market value of that product. The sum of all the market values gives a *valuation of all the market-traded products from the wildlife habitat.*

But this is not all. Consider a housing project in the form of a multi-storeyed building. The windows of the houses open either towards a beautiful forest (or a mountain, or the sea, etc.) or towards a road (or a dump-yard, or a landfill, etc.). A person who lives in an apartment with windows facing the beautiful forest will start everyday admiring the beauty of the forest. On the other hand, a person who lives in an apartment with windows facing the road will start everyday looking at the traffic, and end have nights full of honks. Probably this apartment will also get filled up with the continuous dust and smoke from the vehicles. Given a chance, which apartment would you prefer to buy?

Questions such as these form the basis of the "hedonic pricing method." *Hedonism* is a school of thought that aims at seeking pleasure and avoiding suffering. People go to great lengths to gain pleasure (such as a beautiful sight) and to avoid sufferings (sound, noxious smell, smoke, etc.). Thus, given a chance, people prefer forest- or sea-facing apartments over road- or landfill-facing apartments. This preference can readily be observed through the *differential pricing* of these apartments in the market. Sea-facing apartments typically have a higher price, and fetch a higher value than dump-yard-facing apartments. In such cases, the difference in price that people pay to have an apartment that faces a beautiful forest is an indication of the value that people are putting to the forest. If we added such values as put by different people in the locality, we get an indication of these "pleasure-providing" values of the forest. And this is a value that has been *revealed* through the differences in the market prices.

Similarly, when people come to visit a wildlife habitat such as a tiger reserve, they must pay for several things. There is a cost of planning, a cost of transportation and reaching the tiger reserve, a cost of lodging and boarding, a cost of entry fee, a cost of hiring vehicles and guides, insurance costs, and so on. Since people are paying these costs — and rational thinking would suggest that we never pay above a thing's value — their expenditure can give an indication of the minimum value of the wildlife habitat in their eyes. After all, if one pays ₹20 to buy a pen, the pen must hold a value greater than ₹20 for them. Thus, in the travel cost method, we make a computation of the expenditures that various people make to visit the wildlife habitat. The sum of all of those expenditures is another value of the wildlife habitat, as *revealed* through the price being paid by the visitors.

In the "imputed willingness to pay" methods, we look at circumstantial evidences of the value of the wildlife habitat. One such method is the replacement/substitute cost method. It computes the cost of setting up a substitute for the services being provided by the pristine wildlife habitat. For example, mangrove forests around the sea coasts act as storm and tsunami barriers, providing protection to life and property. If these forests were to be diverted, for the same level of protection, probably a sea wall would have to be constructed. What would be the cost of such a construction? This can be estimated by the dimensions of the wall, the prevailing market prices of raw materials and labour, and the costs of planning, execution and maintenance. Since the mangroves were providing the same service *for free*, they were providing a value that is at least as much as the cost of constructing and maintaining the sea wall.

Similarly, we can compute the costs of setting up substitutes or replacements for the other services being provided by the wildlife habitats. What is the cost of setting up, maintaining and running water purification plants? What is the cost of sewage and waste treatment? What is the cost of erecting geotextile structures to hold soil for preventing soil erosion? What is the cost of desilting dams and waterways? What is the cost of installing air filters to catch dust? Wildlife habitats provide all of these — and more — functions for free, and the monetary value of these *free* services can thus be computed through the replacement/substitute cost method.

Another method is to look at the damage that is being avoided by the well-functioning wildlife habitats. If the wetlands did not purify water, a large number of people living downstream would have suffered gastro-intestinal diseases that are water-borne. What would be the extent of this damage in terms of man days lost and resources diverted to treat them? If mangroves did not protect from storm surges, what value of property would get damaged? If soil filled up the dams, what would be the damage in terms of the reduced life of the dam? Such computations can also provide us with a monetary value of the services being provided by these ecosystems.

It is important to note the difference between methods focussing on revealed versus imputed willingness to pay. When we look at replacement/substitute cost method, we are not *actually* constructing the sea wall or the sewage treatment facility. When we consider damage cost avoided method, we are not actually putting lives and property at risk by destroying the wildlife habitats — to compute the actual losses. What we are doing is a thought experiment — asking "What if?" to get an idea of the monetary value of the ecosystem services. On the other hand, in the revealed willingness to pay methods, we look at the *actual* price being paid in the markets for goods and services. These are not "What if" questions — they actually ask "How much and at what price?" to discern the monetary values. In this way, the "revealed willingness to pay" methods are *computations*, while the "imputed willingness to pay" methods are *estimates*.

In the "expressed willingness to pay" methods, we make use of surveys and questionnaires. In the contingent valuation method, we survey people living near and far about their valuation of the wildlife habitat. To aid the respondents, they may be given hypothetical scenarios such as: "Suppose government wants to cut this forest to raise revenues. But it can also get the revenues through taxation. What is the amount of tax that you are willing to pay to save this forest?" or "How much money are you willing to send to the government of Japan as compensation to persuade them to forego the profit received from hunting whales?" or "Currently you are living in this pristine location near the forest. Suppose you were required to relocate out of this area into a polluted part of the city, full of smoke, noise and smells. What amount of compensation would be a fair compensation for you to relocate from here to there?" Such surveys can also help estimate the value of the wildlife habitats in the eyes of the respondents.

Often a number of methods will be employed concomitantly to get a precise estimate of the monetary value (or the dollar price) of wildlife or their habitats.

11.2.2 Case study of Panna Tiger Reserve

Since wildlife habitats provide a large number of benefits, multiple methods are required for making estimations of different values. Such a system should preferably be GIS-based to permit stratified differential valuations of different parts of the forest or wildlife habitat. One such system is the InVEST model (InVEST = Integrated Valuation of Ecosystem Services and Trade-offs) [Tallis and Polasky, 2009, Nelson et al., 2009, Tallis et al., 2011]. It is a GIS-based suite of open-source software models for mapping and doing valuation of ecosystem services. It performs computations using *spatially explicit data and models*. The final results are obtained in the form of

1. biophysical information (e.g. tonnes of carbon sequestered), and/or

2. economic information (e.g. value of that amount of sequestered carbon).

One study using the InVEST model to compute the value of protected areas [Khanna et al., 2015] employed several means of computing different services, such as

1. employment generation, computed as

$$\Sigma(\text{man days} \times \text{wage rate})$$

2. fishing benefits, computed as

$$\Sigma(\text{production} \times \text{market prices})$$

3. fuelwood benefits, computed as

$$\Sigma(\text{production} \times \text{market prices})$$

4. fodder benefits, computed as

$$\Sigma(\text{production} \times \text{market prices})$$

5. timber benefits, computed as

$$\Sigma(\text{production} \times \text{market prices})$$

6. bamboo benefits, computed as

$$\Sigma(\text{production} \times \text{market prices})$$

7. NTFP benefits, computed as

$$\Sigma(\text{production} \times \text{market prices})$$

8. genepool benefits such as resilience of ecosystems and avenues for future use of biological compounds and other products, computed using *benefits transfer method* — estimating values by transferring available information from other studies in similar location and context.

9. carbon sequestration benefits, computed as

$$\Sigma(\text{sequestration} \times \text{market prices (or using social cost of carbon)})$$

where the *social cost of carbon* is defined as the cost of impacts caused by emission of carbon dioxide,

10. carbon storage benefits, computed as

$$\Sigma(\text{total storage} \times \text{social cost of carbon})$$

11. water provisioning benefits, computed as

$$\Sigma(\text{water provisioned} \times \text{market prices})$$

12. water purification benefits, computed as

$$\Sigma(\text{water purified} \times \text{average cost of treating water (replacement)})$$

13. soil conservation and sediment retention benefits, computed as

$$\Sigma(\text{erosion avoided} \times \text{cost of damage avoided})$$

14. nutrient retention benefits, computed as

$$\Sigma(\text{nutrients retained} \times \text{cost of artificial fertilisers})$$

15. biological control of pests, computed using benefits transfer method,

16. moderation of extreme events benefits, computed using benefits transfer method,

17. pollination benefits, computed using benefits transfer method,

18. nursery for various species benefits, computed using benefits transfer method,

19. habitat for various species benefits, computed using benefits transfer method,

20. cultural heritage benefits, computed using contingent valuation method,

21. recreation benefits, computed using travel cost method,

22. air quality benefits, computed as

$$\Sigma(\text{air purified} \times \text{average cost of treating air (replacement)})$$

23. waste assimilation benefits, computed using benefits transfer method, and

24. climate regulation benefits, computed using benefits transfer method.

For the Panna Tiger Reserve, the various benefits were computed as under:

1. Flow benefits: ₹69.55 billion per year

   (a) Annual direct benefits: ₹0.78 billion
   (b) Annual indirect benefits: ₹53.11 billion
   (c) Option benefits: ₹15.65 billion

2. Stock benefits: ₹137.46 billion per year

3. Critical ecosystem services:

   (a) Water provisioning: ₹25.82 billion per year
   (b) Climate regulation: ₹20.21 billion per year
   (c) Waste assimilation: ₹1.66 billion per year
   (d) Benefits to human health: ₹144.55 billion per year

4. Kinds of services:

   (a) Provisioning services: ₹0.67 billion per year
   (b) Supporting services: ₹0.38 billion per year
   (c) Regulating services: ₹68.48 billion per year
   (d) Cultural services: ₹18.40 million per year

5. **Investment multiplier: 1939.36**

The *investment multiplier* suggests that for every dollar invested in this tiger reserve, the community reaps a benefit of around 2,000 dollars!

The large valuations could suggest to the reader that there are probably some errors in computation. The reality, however, is that well-functioning wildlife habitats *do in fact* provide large amounts of benefits. They are not 'wastelands' waiting to be "put up to good uses" through construction activities. Sadly, we have been *grossly undervaluing* our natural resources in the name of *development*. It is time we took conservation seriously, if only for the benefits it provides to us!

"The forest is a peculiar organism of unlimited kindness and benevolence that makes no demands for its sustenance and extends generously the products of its life activity; it affords protection to all beings, offering shade even to the axe-man who destroys it."

—Gautam Buddha

References

[Akçakaya and Sjögren-Gulve, 2000] Akçakaya, H. R. and Sjögren-Gulve, P. (2000). Population viability analyses in conservation planning: an overview. *Ecological Bulletins*, 48:9–21.

[Awadhiya, 2017] Awadhiya, A. (2017). Impact of global warming on the carbon sequestration potential and stand dynamics of chir pine forests. *Indian Forester*, 143(9):907–914.

[Awadhiya, 2018] Awadhiya, A. (2018). Impact of climate change on wildlife health: A manager's perspective. *Indian Forester*, 144(10):911–921.

[Balmford et al., 1995] Balmford, A., Leader-Williams, N., and Green, M. (1995). Parks or arks: where to conserve threatened mammals? *Biodiversity & Conservation*, 4(6):595–607.

[Beissinger and McCullough, 2002] Beissinger, S. R. and McCullough, D. R. (2002). *Population viability analysis*. University of Chicago Press.

[Berger, 1990] Berger, J. (1990). Persistence of different-sized populations: an empirical assessment of rapid extinctions in bighorn sheep. *Conservation Biology*, 4(1):91–98.

[Blanquet and Pflanzensoziologie, 1964] Blanquet, B. and Pflanzensoziologie, J. (1964). *Grudzuge der Vegetationskunde*. Springer Verlag, Wien, 3rd edition.

[Block and Brennan, 1993] Block, W. M. and Brennan, L. A. (1993). *The habitat concept in ornithology*. Springer.

[Caughley, 1994] Caughley, G. (1994). Directions in conservation biology. *Journal of Animal Ecology*, 63(2):215–244.

[Chen and Chie, 2007] Chen, S.-H. and Chie, B.-T. (2007). *Modularity, product innovation, and consumer satisfaction: An agent-based approach*, pages 1053–1062. Springer.

[Clements, 1916] Clements, F. E. (1916). *Plant succession: an analysis of the development of vegetation*. Number 242. Carnegie Institution of Washington.

[Dangremond et al., 2010] Dangremond, E. M., Pardini, E. A., and Knight, T. M. (2010). Apparent competition with an invasive plant hastens the extinction of an endangered lupine. *Ecology*, 91(8):2261–2271.

[Darwin, 2012] Darwin, C. (2012). *On the Origin of the Species and The Voyage of the Beagle*. Graphic Arts Books.

[Darwin and Bonney, 1889] Darwin, C. and Bonney, T. G. (1889). *The structure and distribution of coral reefs*. Smith, Elder.

[Erwin, 1982] Erwin, T. L. (1982). Tropical forests: their richness in coleoptera and other arthropod species. *The Coleopterists Bulletin*, 36(1):74–75

[Ferguson and Libby, 1971] Ferguson, E. and Libby, W. (1971). Mechanism for the fixation of nitrogen by lightning. *Nature*, 229(5279):37–37.

[Goodman, 1987] Goodman, D. (1987). The demography of chance extinction. *Viable populations for conservation*, 11:34.

[Government, 2020] Government, G. (2020). Press note: Poonam avlokan of asiatic lions in the asiatic lion landscape.

[Hamilton, 1994] Hamilton, M. B. (1994). Ex situ conservation of wild plant species: time to reassess the genetic assumptions and implications of seed banks. *Conservation biology*, 8(1):39–49.

[Hanski and Gilpin, 1991] Hanski, I. and Gilpin, M. (1991). Metapopulation dynamics: brief history and conceptual domain. *Biological Journal of the Linnean Society*, 42(1–2):3–16.

[Heed and Kircher, 1965] Heed, W. B. and Kircher, H. W. (1965). Unique sterol in the ecology and nutrition of drosophila pachea. *Science*, 149(3685):758–761.

[Herbert et al., 1962] Herbert, S. et al. (1962). The architecture of complexity. *Proceedings of the American Philosophical Society*, 106(6):467–482.

[Holt and Bonsall, 2017] Holt, R. D. and Bonsall, M. B. (2017). Apparent competition. *Annual Review of Ecology, Evolution, and Systematics*, 48:447–471.

[Holt and Lawton, 1993] Holt, R. D. and Lawton, J. H. (1993). Apparent competition and enemy-free space in insect host-parasitoid communities. *The American Naturalist*, 142(4):623–645.

[IUCN, 1998] IUCN (1998). *IUCN Guidelines for Re-introductions*. IUCN.

[Janss and Ferrer, 2000] Janss, G. F. and Ferrer, M. (2000). Common crane and great bustard collision with power lines: collision rate and risk exposure. *Wildlife Society Bulletin*, 28(3):675–680.

[Jones, 2001] Jones, J. (2001). Habitat selection studies in avian ecology: a critical review. *The Auk*, 118(2):557–562.

[Khanna et al., 2015] Khanna, C., Singh, R., David, A., Edgaonkar, A., Negandhi, D., Verma, M., Costanza, R., and Kadekodi, G. (2015). *Economic valuation of tiger reserves in India: A Value+ Approach*. Indian Institute of Forest Management.

[Kitching and Ebling, 1961] Kitching, J. and Ebling, F. (1961). The ecology of lough ine. *The Journal of Animal Ecology*, 30(2):373–383.

[Klopfer, 1963] Klopfer, P. (1963). Behavioral aspects of habitat selection: the role of early experience. *The Wilson Bulletin*, 75(1):15–22.

[Kole et al., 2011] Kole, R., Karmakar, P., Poi, R., Mazumdar, D., et al. (2011). Allelopathic inhibition of teak leaf extract: A potential pre-emergent herbicide. *Journal of Crop and Weed*, 7(1):101–109.

[Lacy, 1993] Lacy, R. C. (1993). Vortex: a computer simulation model for population viability analysis. *Wildlife Research*, 20(1):45–65.

[Lande and Barrowclough, 1987] Lande, R. and Barrowclough, G. (1987). *Effective population size, genetic variation, and their use in population*. chapter 6, page 87. Cambridge University Press.

[Leopold, 1933] Leopold, A. (1933). Game management. 481 pp., illus. *New York*.

[Lindenmayer et al., 1995] Lindenmayer, D. B., Burgman, M., Akçakaya, H., Lacy, R., and Possingham, H. (1995). A review of the generic computer programs alex, ramas/space and vortex for modelling the viability of wildlife metapopulations. *Ecological Modelling*, 82(2):161–174.

[Macias et al., 2000] Macias, F., Lacret, R., Varela, R., and Nogueiras, C. (2000). Allelopathic potential of teak (tectona grandis). In *Allelopathy-From Understanding to Application. Proceedings of Second European Alleopathy Symposium*, page 140. Citeseer.

[Madsen et al., 1999] Madsen, T., Shine, R., Olsson, M., and Wittzell, H. (1999). Restoration of an inbred adder population. *Nature*, 402(6757):34–35.

[Madsen et al., 2004] Madsen, T., Ujvari, B., and Olsson, M. (2004). Novel genes continue to enhance population growth in adders (vipera berus). *Biological Conservation*, 120(1):145–147.

[Mayr, 1942] Mayr, E. (1942). *Systematics and the origin of species*. Columbia University Press.

[Meena, 2009] Meena, V. (2009). Variation in social organisation of lions with particular reference to the asiatic lions panthera leo persica (carnivora: Felidae) of the gir forest, india. *Journal of Threatened Taxa*, 1(3):158–165.

[Meiners, 2007] Meiners, S. J. (2007). Apparent competition: an impact of exotic shrub invasion on tree regeneration. *Biological Invasions*, 9(7):849–855.

[Mlot, 2013] Mlot, C. (2013). Are isle royale's wolves chasing extinction? *Science*, 340(6135):919–921.

[Mlot, 2015] Mlot, C. (2015). Inbred wolf population on isle royale collapses. *Science*, 348(6233):383.

[Monteith, 1972] Monteith, J. (1972). Solar radiation and productivity in tropical ecosystems. *Journal of Applied Ecology*, 9(3):747–766.

[NASA, 2003] NASA (2003). Image of the day for april 22, 2003. Webpage.

[NASA, 2005] NASA (2005). Svs: Rondonia deforestation (wms). Webpage.

[NASA, 2009] NASA (2009). Image of the day for june 24, 2009. Webpage.

[NASA, 2012] NASA (2012). Measuring carbon and trees in the tropics.

[NASA, 2017] NASA (2017). Below the mangrove canopy.

[NASA, 2018a] NASA (2018a). Landsat image gallery — a new reservoir in cambodia. Webpage.

[NASA, 2018b] NASA (2018b). Landsat image gallery — brazil's carajas mines. Webpage.

[NASA, 2019a] NASA (2019a). Gedi mission early success.

[NASA, 2019b] NASA (2019b). Return of the gedi's first data.

[NASA, 2020a] NASA (2020a). Earth in the third dimension: First gedi data available.

[NASA, 2020b] NASA (2020b). A new measuring stick for forests.

[Nelson et al., 2009] Nelson, E., Mendoza, G., Regetz, J., Polasky, S., Tallis, H., Cameron, D., Chan, K. M., Daily, G. C., Goldstein, J., Kareiva, P. M., et al. (2009). Modeling multiple ecosystem services, biodiversity conservation, commodity production, and tradeoffs at landscape scales. *Frontiers in Ecology and the Environment*, 7(1):4–11.

[Newbold et al., 2016] Newbold, T., Hudson, L. N., Arnell, A. P., Contu, S., De Palma, A., Ferrier, S., Hill, S. L., Hoskins, A. J., Lysenko, I., Phillips, H. R., et al. (2016). Has land use pushed terrestrial biodiversity beyond the planetary boundary? a global assessment. *Science*, 353(6296):288–291.

[NOAA, 2020] NOAA (2020). Impacts — or&r's marine debris program.

[O'Brien et al., 1985] O'Brien, S. J., Roelke, M. E., Marker, L., Newman, A., Winkler, C., Meltzer, D., Colly, L., Evermann, J., Bush, M., and Wildt, D. E. (1985). Genetic basis for species vulnerability in the cheetah. *Science*, 227(4693):1428–1434.

[O'Brien et al., 1987] O'Brien, S. J., Wildt, D. E., Bush, M., Caro, T. M., FitzGibbon, C., Aggundey, I., and Leakey, R. E. (1987). East african cheetahs: evidence for two population bottlenecks? *Proceedings of the National Academy of Sciences*, 84(2):508–511.

[O'Brien et al., 1983] O'Brien, S. J., Wildt, D. E., Goldman, D., Merril, C. R., and Bush, M. (1983). The cheetah is depauperate in genetic variation. *Science*, 221(4609):459–462.

[Orians and Collier, 1963] Orians, G. H. and Collier, G. (1963). Competition and blackbird social systems. *Evolution*, 17(4):449–459.

[Parker, 1963] Parker, J. (1963). Cold resistance in woody plants. *The Botanical Review*, 29(2):123–201.

[Possingham et al., 2013] Possingham, H. P., McCarthy, M. A., and Lindenmayer, D. B. (2013). Population viability analysis. In Levin, S., editor, *Encyclopedia of Biodiversity (Second Edition)*. Elsevier.

[Power et al., 1996] Power, M. E., Tilman, D., Estes, J. A., Menge, B. A., Bond, W. J., Mills, L. S., Daily, G., Castilla, J. C., Lubchenco, J., and Paine, R. T. (1996). Challenges in the quest for keystones: identifying keystone species is difficult—but essential to understanding how loss of species will affect ecosystems. *BioScience*, 46(8):609–620.

[Raab et al., 2011] Raab, R., Spakovszky, P., Julius, E., Schuetz, C., and Schulze, C. H. (2011). Effects of power lines on flight behaviour of the west-pannonian great bustard *Otis tarda* population. *Bird Conservation International*, 21(2):142–155.

[Räikkönen et al., 2009] Räikkönen, J., Vucetich, J. A., Peterson, R. O., and Nelson, M. P. (2009). Congenital bone deformities and the inbred wolves (*Canis lupus*) of isle royale. *Biological Conservation*, 142(5):1025–1031.

[Rowland and Vojta, 2013] Rowland, M. M. and Vojta, C. D. (2013). *A technical guide for monitoring wildlife habitat*. USDA.

[Shaffer, 1981] Shaffer, M. L. (1981). Minimum population sizes for species conservation. *BioScience*, 31(2):131–134.

[Silva et al., 2014] Silva, J. P., Palmeirim, J. M., Alcazar, R., Correia, R., Delgado, A., and Moreira, F. (2014). A spatially explicit approach to assess the collision risk between birds and overhead power lines: a case study with the little bustard. *Biological Conservation*, 170:256–263.

[Siraj et al., 2014] Siraj, A., Santos-Vega, M., Bouma, M., Yadeta, D., Carrascal, D. R., and Pascual, M. (2014). Altitudinal changes in malaria incidence in highlands of ethiopia and colombia. *Science*, 343(6175):1154–1158.

[Stevenson, 2010] Stevenson, A. (2010). *Oxford dictionary of English*. Oxford University Press, USA.

[Tallis and Polasky, 2009] Tallis, H. and Polasky, S. (2009). Mapping and valuing ecosystem services as an approach for conservation and natural-resource management. *Annals of the New York Academy of Sciences*, 1162(1):265–283.

[Tallis et al., 2011] Tallis, H., Ricketts, T., Guerry, A., Nelson, E., Ennaanay, D., Wolny, S., Olwero, N., Vigerstol, K., Pennington, D., Mendoza, G., et al. (2011). *InVEST 2.1 beta user's guide*. The Natural Capital Project, Stanford University, University of Minnesota, The Nature Conservancy, and World Wildlife Fund.

[Turček and Hickey, 1951] Turček, F. and Hickey, J. (1951). Effect of introductions on two game populations in czechoslovakia. *The Journal of Wildlife Management*, 15(1):113–114.

[Wayne et al., 1991] Wayne, R., Lehman, N., Girman, D., Gogan, P., Gilbert, D., Hansen, K., Peterson, R., Seal, U., Eisenhawer, A., Mech, L., et al. (1991). Conservation genetics of the endangered isle royale gray wolf. *Conservation Biology*, 5(1):41–51.

[Wildt et al., 1987] Wildt, D. E., Bush, M., Goodrowe, K., Packer, C., Pusey, A., Brown, J., Joslin, P., and O'Brien, S. J. (1987). Reproductive and genetic consequences of founding isolated lion populations. *Nature*, 329(6137):328–331.

[Wildt et al., 1983] Wildt, D. E., Bush, M., Howard, J., O'Brien, S. J., Meltzer, D., Van Dyk, A., Ebedes, H., and Brand, D. (1983). Unique seminal quality in the south african cheetah and a comparative evaluation in the domestic cat. *Biology of reproduction*, 29(4):1019–1025.

[Wilson and MacArthur, 1967] Wilson, E. O. and MacArthur, R. H. (1967). *The theory of island biogeography*. Princeton University Press.

Index